P9-DTN-347

3/13/12
$75.00

Investing in the Renewable Power Market

Founded in 1807, John Wiley & Sons is the oldest independent publishing company in the United States. With offices in North America, Europe, Australia and Asia, Wiley is globally committed to developing and marketing print and electronic products and services for our customers' professional and personal knowledge and understanding.

The Wiley Finance series contains books written specifically for finance and investment professionals as well as sophisticated individual investors and their financial advisors. Book topics range from portfolio management to e-commerce, risk management, financial engineering, valuation and financial instrument analysis, as well as much more.

For a list of available titles, visit our Web site at www.WileyFinance.com.

Investing in the Renewable Power Market

*How to Profit from
Energy Transformation*

TOM FOGARTY

AND

ROBERT LAMB

WILEY

John Wiley & Sons, Inc.

Published by John Wiley & Sons, Inc., Hoboken, New Jersey.
Published simultaneously in Canada.

For general information on our other products and services or for technical support, please
contact our Customer Care Department within the United States at (800) 762-2974, outside
the United States at (317) 572-3993 or fax (317) 572-4002.

Wiley also publishes its books in a variety of electronic formats. Some content that appears in
print may not be available in electronic books. For more information about Wiley products,
visit our web site at www.wiley.com.

Library of Congress Cataloging-in-Publication Data:

Fogarty, Tom, 1963—
    Investing in the renewable power market : how to profit from energy transformation /
Tom Fogarty and Robert Lamb.
        p. cm. — (Wiley finance ; 614)
    Includes bibliographical references and index.
    ISBN 978-0-470-87826-2 (cloth); ISBN 978-1-118-22102-0 (ebk);
    ISBN 978-1-118-23478-5 (ebk); ISBN 978-1-118-25936-8 (ebk)
    1.   Renewable energy resources–United States–Finance. 2.   Investments–United States.
I. Lamb, Robert, 1941- II. Title.
    TJ807.9.U6 F64 2012
    333.79'40681—dc23   2011043312

10 9 8 7 6 5 4 3 2 1

To Yayoi and Atticus

Contents

Acknowledgments

Tom would like to thank his wife Yayoi, who encouraged him to write a book about the challenges of renewable and fossil energy project development. Tom's thinking about the energy business has been influenced by a number of people he has worked with. This includes Roy Cuny, Larry Grundmann, Richard Grosdidier, Phil Burkhardt and Seth Arnold. He was fortunate to learn about distressed and entrepreneurial investing from Art Rosenbloom, MaryJane Boland, Ed Altman, Roy Smith, and Allan Brown at NYU Stern.

Bob and Tom truly appreciate all of the efforts in creating and organizing this book from Debra Englander, Jennifer MacDonald, and Donna Martone of Wiley. Their work was always professional and they were instrumental in reviewing concepts.

T.F and R. L.

Introduction

This book was written to help investors, energy practitioners, and students understand the limits of renewable power. It is intended to help the reader learn how to evaluate renewable power investments by reviewing the technical and economic issues for fossil power as well as wind, solar, thermal, and other renewable power technologies.

Together, we bring both an academic and a practitioner perspective to this book. Tom is an energy executive at a major international energy corporation. He was a financial-management and energy consultant and a former executive MBA student of Professor Bob Lamb. Bob has published books and chapters on energy, finance, and strategic management. He has been debt adviser to New York Power Authority, the U.S. government, various states, public authorities, and corporations. Bob has also served as expert witness in litigations and arbitrations. Tom and Bob have jointly collaborated on a number of energy projects. Both of us, in our work and experience, were frustrated that there was not a text available to soberly evaluate renewable power investments in today's complex energy environment.

The global energy markets have never been so complex and fast changing. The United States has recently discovered very large amounts of shale gas embedded in solid rock formations extending from New York to Texas and California. That shale gas is extractable via new technologies. There is a concern that this plentiful and currently cheap natural gas could make the United States complacent about its energy future. The development and use of hydraulic fracturing has made shale gas an energy game changer. This new natural gas supply makes the overall economics difficult for renewable energies and for coal and nuclear power plants.

In fact, the solar company Solyndra filed for bankruptcy at the end of August 2011 despite having received a 2009 cash grant of $535 million from the U.S. Treasury. Beacon Power, a manufacturer of flywheel based energy storage systems also filed for bankruptcy on October 31, 2011. The extensive U.S. natural gas pipeline and storage infrastructure have made shale gas an immediate player in the energy marketplace. Other countries are also discovering shale gas supplies but might not have the infrastructure to distribute it to end users.

Even more important, these national, political, and social conflicts have been taking place simultaneously with the rapid pace of several major new technology changes and broad, intense global ramp-ups of various types of unconventional energy, gas, shale gas, tar sands extraction, innovations at ultra-deep horizontal wells. Oil and gas drilling innovations in semi-submersibles and arctic drilling platforms.

The major Japanese earthquake, tsunami, and the Fukushima Daiichi nuclear power plant "meltdown," along with Germany's decision to close its nuclear reactors, are deeply impacting the worldwide nuclear power industry. In the United States, the Indian Point nuclear power plant operating license extension is being challenged by various New York state agencies. Their concern is that the facility is too close to New York City.

This book is meant to be a current, realistic analysis of the various changes in the development of renewable energy technologies and how these technologies compare to fossil energy production, energy storage technologies, and energy transmission and demand-side management. One of the most important points of this book is to stress the essential need for multiple types of energy plus coordination mechanisms across both nations and continents. Countries continue to take an approach of either all coal or all nuclear and, more recently, all natural gas. We will need all energy sources in the future.

World scientists now appear to be in agreement that global warming is not solely in evidence in the melting of the polar ice caps but is in evidence in the radical changes in global weather conditions. Little progress has been made despite more than a decade of intense international conferences and pledges of global support for concerted efforts to finally cope with the risks and very clear dangers of global climate changes. They are concerned about record winter snowstorms, record spring rains, record flooding, record droughts and famines, and the escalation of record number of global earthquakes.

Another vital energy and health challenge is that there exists no available technology to economically remove and sequester CO_2 at scale. This point never seems to be made during discussions of climate change.

Due to major budget deficits, the United States and Europe have now been or will be forced to drastically cut back major expenditures on their "various green energy initiatives." The U.S. "1603 cash grant for renewable power plants" will finish at the end of 2011 and is unlikely to be extended in the current political environment.

U.S. states continue to be unrealistic on the size of their renewable energy portfolio standards. Some have gone so far as to count on renewable facilities that would be built in other states. China is currently willing to subsidize the production of solar power plant technology. This will force the United States to continue to be an innovator and to look at China as a partner and not a competitor.

Economic downturns in Spain forced the government to cancel direct government subsidies plus energy grants for solar panels and wind farms. Eliminating energy feed-in tariffs would leave many major energy companies with losses. It raises important questions about whether most new green energy programs generally will be a victim of the long-term U.S. and European recession economies.

Let's start with Chapter 1, where we provide the reader with the fundamentals of evaluating renewable power.

An Overview of Renewable Power

The Walrus and the Carpenter were walking close at hand; They wept like anything to see Such quantities of sand: "If this were only cleared away," They said, "it would be grand!"

"If seven maids with seven mops Swept it for half a year. Do you suppose," the Walrus said, "That they could get it clear?" "I doubt it," said the Carpenter, And shed a bitter tear.
— from *Through the Looking Glass*, Lewis Carroll

Wind, solar, and geothermal renewable power technologies face a number of technological challenges. A typical wind power project has yearly availability only in the low 30 percent range and a typical solar photovoltaic project has an availability of approximately 16 percent. For further clarification, a 100–megawatt (mW) wind project could produce only 262,800 megawatt hours (mWh) in one year (e.g., 100 mW × 365 days/yr × 24 hours/day × 30 percent = 262,800 mWh). Solar thermal projects have a higher availability but are more expensive, have regulatory permitting challenges, and are typically not located in liquid power trading markets.

Electricity, unlike other commodities, can't be stored, which leads to a large amount of volatility in electricity prices. It is important to remember that current battery technology is only capable of storing electricity for up to four hours. Switching the world's energy supply to renewable power is not like starting the next Google. It is not a case of placing five or six smart boys and girls in a room and asking them to think up the next clean energy technology. Older facilities are endowed with a scarcity value due to the difficulty of obtaining air permits for new fossil power plants. These are a number of the issues that make it difficult for renewable power plants to be competitive with traditional fossil power plants.

The economic profile of a typical wind and solar power project is a small amount of earnings before interest, taxes, depreciation, and amortization (EBITDA), tax credits, and accelerated depreciation. Only EBITDA can be used to pay down debt, and overestimating kilowatt hours (kWh) produced or underestimating maintenance expense can lead to a debt default. Since there is currently no inexpensive, high-capacity battery technology, it is not possible to run the electric grid with wind and solar power on a 24/7 basis. For the next few years, the economics for these types of projects will work when constructed projects can be purchased on a distressed basis for cents on the dollar. An October 16, 2009, *Forbes* article stated that the U.S. Energy Information Administration estimates that a kilowatt hour of electricity from a photovoltaic (PV) solar plant entering service in 2016 will cost 40 cents/ kWh in 2009 dollars. The article further stated that this is three to five times the projected cost of electricity generated from natural gas, coal, or uranium.[1]

Other than battery technology, the only way to firm up the power produced from these facilities is to use gas turbine engines. One study under development will show that a gas turbine engine that provides backup to a wind power plant actually produced more carbon dioxide (CO_2) emissions than a coal-fired power plant. A gas turbine engine would also require a local supply of natural gas and pipeline transportation, which might not be available. Improvements in gas turbine engine design might also be required to meet both quick-start and a low-emission profile. There is also a shortage of electric transmission in locations where there is potential for wind power projects.

IT'S ALL ABOUT NATURAL GAS

A key challenge is that prices for renewable power in the United States are priced off of natural gas, which is currently at historic lows. Recent large unconventional natural gas finds throughout the United States should continue to support natural gas prices below $7 per million British Thermal Units (MMBtu).

The Energy Information Administration estimates that the United States has approximately 1,770 trillion cubic feet (Tcf) of technically recoverable gas, including 238 Tcf of proven reserves. The Potential Gas Committee estimates total U.S. gas resources at 2,074 Tcf. It is estimated that technically recoverable unconventional gas including shale accounts for nearly two-thirds of American onshore gas resources. At the current production rates, "the current

---

[1] Duncan Greenberg, "Seeing the Light," *Forbes*, October 19, 2009.

recoverable resource estimate provides enough natural gas to supply the United States for the next 90 years."[2] In 1996, it was estimated that the Barnett Shale contained only 3 Tcf of reserves. As a result of an upgrade in technology, it was estimated in 2006 that it now contains 39 Tcf. The natural gas was always there; it was just not possible to get to it with older technology. At the time of this writing, this new supply of natural gas has resulted in natural gas prices in the $4/MMBtu price range. This is a large drop from $13.57/MMBtu in the summer of 2008. The economics for natural gas storage projects still appear to work since there has been weekly and seasonal movement in price.

This unconventional natural gas supply situation has resulted in the return on equity for renewable power plants to be actually lower than the return from purchasing a secured loan in some existing natural gas–fired power producers. That an equity security in the bottom of the capital structure of a yet-to-be-constructed power plant could require a lower return than a first lien secured loan in an operating gas-fired power plant doesn't make sense. This situation is similar to the real estate crisis, whereby real estate market mortgage loans written prior to 2007 were priced at levels that didn't reflect their risk and ultimately defaulted. This situation means that in the short term it makes sense to bet against renewables as opposed to developing or investing in new projects. By comparison, renewable power producers in Europe are currently paid a very high fixed price for power under the national government's feed-in tariff program. This situation is also unsustainable. A May 20, 2010, *Bloomberg* article entitled "Greek Crisis and Euro's Drop Snare Clean-Energy Stocks" stated:

> . . . *The aid to renewable energy, paid by consumers through their power bills, is being slashed by governments aiming to curb their own budget deficits and to cut energy costs for businesses and consumers. . . .*[3]

CONTROL OF CO₂ EMISSIONS IS NOT CURRENTLY POSSIBLE

Another challenge that renewable power faces is that there is currently no proven technology to remove CO_2 emissions from existing power plants. This is a critical fact that makes it difficult to switch from traditional fossil

---

[2] John D. Podesta and Timothy E. Wirth, *Natural Gas: A Bridge Fuel for the 21st Century* (Washington, DC: Center for American Progress, August 10, 2009).
[3] Ben Sills and Mark Scott, "Greek Crisis and Euro's Drop Snare Clean-Energy Stocks," *Bloomberg*, May 20, 2010.

power plants to renewable power plants and has helped to make it difficult for a carbon cap and trade or tax to pass. Carbon emissions not only have to be removed from the stack of a power plant but also have to be sequestered or buried deep underground. This task requires a large amount of energy and a corresponding increase in production cost. The key pollutants that existing fossil power plants (e.g., coal and natural gas) emit include sulfur dioxide (SO_2), nitrogen oxide (NOx) and particulates (in the case of coal power plants). With current technology, it is relatively easy to control these pollutants. In order to control SO_2, a scrubber is used; in the case of NOx, selective catalytic reduction (SCR) is used; and for particulate, an electro-static participator is used.

CO_2 scrubbers are currently being tested by the French technology company Alstom at the electric utility American Electric Power. The CO_2 cost resulting from the proposed congressional cap-and-trade program is based on mostly free or low-cost allowances. As a result, it will be cheaper to use allowances as opposed to purchasing an unproven, expensive emission control technology. There is a strong possibility that Congress will not pass CO_2 legislation, and this task will fall to the Environmental Protection Agency (EPA). The EPA will then use its New Source Performance Standard (NSPS) to determine the best available control technology (BACT) for CO_2. Since there is no "available" technology to control CO_2, the EPA would have a difficult time regulating CO_2. With pollutants such as oxides of nitrogen, a power plant can either buy allowances or install a proven control technology. This will not help the overall economics for renewable power projects.

The EPA will have a difficult time attempting to regulate CO_2 under its existing BACT program. In past BACT rulings, if a technology was not commercially available or too expensive for pollutants such as NOx, a power plant would not be required to include this technology in its design. This is also true for operating power plants that fall under reasonably available control technology (RACT). Alstom has stated in a recent *Financial Times* article that their CO_2 control technology for coal-fired power plants will not be ready until 2015. It is quite possible that even this date is too optimistic, and a generator could claim that this technology is still not commercially available.[4]

Cogeneration as a CO_2 Control Technology

The existing BACT and RACT regulation will also not allow the EPA to force generators to change their fuel supply or their initial technology

---

[4] Anna Fifield, "Uncertainty over Emission Targets Hinders Investment," *Financial Times*, October 24, 2009.

selection in order to reduce CO_2 emission. As a result, a coal-fired power plant could not be required to cofire biomass. In new air permits, the generator proposes its fuel supply. If a generator wants to change its fuel supply, it has to modify its existing permit and conduct additional air modeling studies.

If CO_2 control technology were commercially available and cost, for example, \$100/ton per ton of CO_2 removed, the EPA could mandate that this was BACT/RACT. Generators could then make a cost argument unless the EPA ruled that CO_2 fell under lowest achievable emission reduction (LAER). Under LAER, generators can't argue that a particular technology is too expensive to install. Under both LAER and BACT/RACT, a technology still has to be commercially available. In order to accomplish this, the EPA would have to argue that each region of the United States is nonattainment or exceeds federal standards for CO_2 emissions. This would be difficult for the EPA to do since it has already ruled that CO_2 emissions are a global problem and not a regional one. To regulate CO_2 emissions, the fastest approach continues to be cap and trade or carbon tax regulation.

In the short term, combined heat and power (CHP) or cogeneration power projects will provide a bridge to reducing CO_2 emissions. CHP power plants are located at the site of an industrial steam or process heat user or at a district heating/cooling system and provide the ability to work one fuel twice. A CHP power plant would allow an industrial factory to run its existing boilers on standby since it would produce both steam and power. CHP typically allows an industrial factory to produce power at a lower cost than it can buy from the grid and to produce heat at a lower cost than from its existing boilers. According to the EPA, a typical CHP power plant can have an overall cycle efficiency of 75 percent, while conventional generation has only a 49 percent overall efficiency.

CHP can also be a source of electric power in a congested area where it might not otherwise be possible to site a traditional power plant. Depending on the size of the CHP plant and the amount of steam sold, this can result in a drop in not only CO_2 emission but also NOx, SO_2, CO, and other criteria air pollutants. A typical CHP power plant employs efficient gas turbine engine technology, which is lower in emission output than the local power plants. The states of New York and New Jersey have realized how CHP can cut air emissions and provide power in critical and/or transmission-constrained locations. As a result, they have been providing financial incentives for the development of CHP power plants. One New Jersey regulator recently stated that CHP projects should be supported since solar power is only 16 percent available and is very expensive.

REALITY OF DEMAND-SIDE MANAGEMENT

Utilities are currently implementing a number of smart grid and energy efficiency or demand-side management programs. These programs include encouraging industrial consumers to shift load to off-peak times and installing smart meters, which provide real-time power pricing and communication with the central office of an electric utility. Programs of this type could help reduce the need to build new power plants and/or reduce the operation of existing power plants. Peaking power plants would be most affected by an increase of demand-side management. Peaking power plants are the highest emitters of air pollution, so this would result in a subsequent reduction in CO_2 and other air pollutants.

The concern is that future demand-side management reduction could be overestimated since industrial customers might not reduce their load when it affects their own customer's requirements. An example would be a hotel not turning down the air conditioning on a hot day when power supplies are tight so as not to upset its guests. Some residents might not be able to live in an apartment building that turns its common-area lighting off when not in use due to religious customs. Since most residential customers pay an average price for power and are not currently exposed to real-time electricity pricing, their bills under a real-time pricing program could actually increase in the future. Consumers would not be willing to have their high-energy-consuming televisions turned off during their favorite program in order to reduce load. Certain industrial customers have a number of different options to shift load to off-peak times. Residential customers can run the dishwasher and do the laundry only in the evening.

Apartment buildings have converted clothes dryers from electric to natural gas. This helps to increase natural gas load during a low-gas-consumption period in the summer and to reduce electric power load during a high-consumption period in the summer. Natural gas dryers have a higher initial cost and are more efficient than electric dryers. This initial cost can be offset by a grant to the apartment building by the local natural gas and electric utility, which both benefit. However, this will not result in a material amount of reduction in load. This situation has made it difficult for smart metering and energy efficiency programs to be approved by public service commissions. Due to legacy liability issues with existing power meters, there is a large cost to switch to smart meters. In most cases, regulated utilities would bear this cost and be required to raise rates to accomplish this.

Unlike large coal, nuclear, and natural gas power projects, wind and solar power projects will not create a large amount of "green collar" jobs. Photovoltaic solar projects that are under development are typically only in the 1- to 2-mW size range and have a very small permanent staff. This staffing issue is also true for wind turbine power plants. Even the require-ment to weatherize and relamp existing buildings will not create a large number of permanent full-time jobs.

SUMMARY

Renewable power plants compete in a world of operating fossil power plants and currently inexpensive natural gas. Despite many governments financial grants, subsidies, and tax incentives, a large number of alternative energy companies have failed over the past 30 years. In the short term, investment opportunities with renewable power will be with financial and operation restructurings of troubled or bankrupt power plants. Longer term, competitively priced battery storage will have to be developed, along with a very high price for CO_2 emissions, in order to support the growth of renewable power. In any event, power investors will want to continue to monitor investing in renewable power in order to avoid the mistake that the old AT&T made by continuing to focus on long distance fixed-line telephone business and ignoring the mobile telephone market.

Chapter 2 describes how to accomplish this task.

Analyzing Power Project Economics

If you give a man a fish, you feed him for a day. If you teach a man to fish, you feed him for a lifetime.

—Chinese proverb

Regardless of the technology used by a particular power plant, the overall economics have to exceed its risk-adjusted cost of capital. As previously mentioned, investors must also consider relative value and their position in the capital structure of a particular investment.

REGULATED UTILITIES

Fossil or renewable power projects can be owned by an independent power producer (IPP) or placed in rate base by a regulated utility. Regulated utilities are granted a service territory, where they are allowed to earn a return on generation, transmission, and distribution investment and maintenance. *Generation* refers to the actual production of electricity; *transmission*, the movement of electricity from the power plant over transmission lines to the end user; and *distribution*, final delivery to the end user. In some markets, regulated utilities have been forced to sell off their generation and not allowed to build any new power plants. In this type of market, the regulated utility can earn a return only on new and existing transmission and distribution.

The concept for regulated utilities is that if the utility spends $1 on fuel or operations and maintenance, the utility recovers $1 in rates. A typical utility has a regulated capital structure that is 50 percent debt,

45 percent equity, and five percent preferred stock. Assuming that the debt had a cost of six percent, the equity 10.5 percent, and the preferred stock 11.5 percent, the utility would have an after-tax cost of capital of 7.10 percent, based on a 40 percent combined federal and state tax rate. A former boss used to say to me that a regulated utility goes out of business each year! What he meant by this is that the regulated rate base of a regulated utility is reduced each year by depreciation, and utilities have an incentive to build new power plants or make other types of investments that can go into the rate base. With a regulated rate base, some of the issues with variability in power production of renewable power go away. This is due to the fact that with a regulated rate base, operating costs are a pass-through and there is a regulated return on capital no matter how many hours the power plant operates. However, the issue of grid stability and reliability doesn't go away.

We have heard, from a recent conversation with an airline employee, that airline firms have traditionally not earned above their cost of capital. In the past, airlines got around this issue by having a regulated rate base with a guaranteed return. This approach could work for renewable power plants as long as public service commissioners did not place limits on overall capital and or operating cost. A regulated utility is also required to show that they have a need for future capacity. The best way to understand the concept of cost of capital is to consider taking a cash advance on a credit card at, say, an 18 percent interest rate and then investing this cash in a savings account paying 3 percent. This transaction results in a loss of 15 percent. Of course, this type of capital structure and business model is not sustainable.

More recently, regulated utilities have been given an incentive by their regulators to earn a return for selling less power. An example of this could occur if a utility installed a control to limit the output of a central air conditioner in a home on a hot summer day. If enough of these devices were installed, the utility could reduce the output or turn off one or more power plants. Since each air conditioner would run for fewer hours, the utility would produce less megawatt hours of electricity. Under a traditional rate-making process, a project of this type would result in a loss of revenue for the utility. In addition to a reduction in air emissions, this approach could also avoid the installation of new power plants in the future. This demand side as opposed to supply side (e.g., addition of a new power plant) approach results in a reduction in overall power consumption.

The simple revenue requirement for a typical gas turbine combined-cycle (GTCC) power plant assuming a 50 percent debt, 45 percent equity, and five percent preferred stock capital structure typical of a regulated utility is shown in Table 2.1.

TABLE 2.1 Gas Turbine Combined-Cycle (GTCC) Revenue Requirement

Capacity	530	MW	Fixed to Variable Cost Conversion
Capital cost	$530,000	$(000)	
$/kW	1,000		
Capacity requirement—$/kW-yr	$ 95		
Capacity requirement-$/kW-month	$ 7.93		50,418
Fixed O&M—$/kW-month	$ 1.89		12,020
Capacity and fixed—$/kW-month	$ 9.82		
Variable O&M—$/MWh	$ 2.83		
Energy price—$/MWh	$ 43		
Natural gas price—Delivered $/MMBtu	$ 6		
Heat rate—Btu/KWh	7,200		
Assumed dispatch	60%		
MWh production	2,785,680		$22.41/MWh
Strike Price—Var O&M + Energy—$/MWh	$ 46		
All-in total cost ($/MWh) @ assumed dispatch	$ 68.44		

Capital Structure

Debt	6.00%
Debt tax effect—40% Rate	3.60%
Debt percentage	50.00%
Preferred stock	11.50%
Preferred stock percentage	5.00%
Equity—A/T	10.50%
Equity	45.00%
Weighted average cost of capital @ (50/5/45)	7.10%
Investment term	20 Years

The analysis is based on an all in capital cost/enterprise value of $530,000,000 or $1,000/kW for a 530-mW power plant. This project is typical of current and future power projects that will be added to the U.S. power grid. Actual project costs can vary greatly and will depend on local conditions and the level of liquidated damages. Fixed operating and maintenance cost mostly covers the plant staffing cost. Variable operating and maintenance covers both maintenance and chemical costs. The plant is assumed to operate 60 percent of the year and produces 2,785,680 mWh (530 mWh/hour × 8,760 hours/year × 60 percent = 2,785,680 mWh). The price for natural gas includes both the natural gas itself and pipeline transportation to the plant.

The preceding 7-day-a-week, 24-hour-a-day total cost or revenue requirement of $68.44 would be compared with forward pricing quotes from Bloomberg and brokers to assess overall project economics. In order to determine the long-term value of a power plant, forward calendar prices are used. The typical calendar price is based on a 5-day-a-week, 16-hour-a-day schedule. For the entire year, this is a total of 4,160 hours, which is based on 5 days/week × 16 hours/day × 52 weeks/year. This is defined as 5 × 16 or an on-peak price during daytime hours, which the plant would obtain for 47.49 percent of the year. This on-peak time is the period when there is the largest requirement of power, and as a result power is most valuable. During the remaining 52.51 percent of the year (or 1 to 47.49 percent) the plant would receive the off-peak price for power. Note that the Bloomberg service has power pricing for a number of the major U.S. electric and natural gas trading hubs.

The pricing information available through Bloomberg helps an investor determine if his plant will dispatch or operate in the forward market and his future earnings before interest, taxes, depreciation, and amortization (EBITDA). Bloomberg also has off-peak pricing, which can be used to calculate a 7 × 24 or around-the-clock (ATC) power price. The market for capacity is not as liquid, is region/independent system operator (ISO) specific, and is not quoted by Bloomberg. A capacity payment can be considered an insurance payment. In the case of a peaking power plant, there may be only a few times a year when it is needed to operate. The capacity payment is meant to cover the return to debt, equity, and fixed operating and maintenance cost for this plant when it is not operating. Some markets are quoting capacity prices for only three years and others only by summer and winter season. This lack of price discovery tends to discourage the building of new power plants since it can take two years to develop and obtain unappealable air permits and another two years to fully construct and start up a natural gas combined-cycle power plant. As a result, a power project developer would not know what his price for capacity would be after he had started final development of his project and would not be able to obtain long-term, low-cost, nonrecourse project finance debt. At current power prices, an investor without a rate base can't justify building a new combined-cycle power plant.

EVALUATING A POWER PLANT

When evaluating a power plant, an important first step is to calculate its variable cost. The variable cost for a fossil-based power plant includes its fuel and variable operating maintenance cost. Fuel cost is priced in

$/MMBtu and is translated into an energy cost priced in $/mWh or $/kWh by multiplying heat rate by gas cost. Heat rate is specified in Btu/kWh, and the lower the number, the more efficient the power plant is. A typical natural gas combined-cycle power plant will have a heat rate of 7,200 Btu/kWh, while a simple-cycle gas plant would have a heat rate of 10,000 Btu/kWh to 11,000 Btu/kWh, and a coal plant would have a heat rate from 8,800 Btu/kWh to 13,000 Btu/kWh depending on heat rate and coal quality. Variable operating and maintenance costs typically include chemicals, consumables, and costs related to hours of plant operation. A typical 500-mW natural gas–fired power plant with a heat rate of 7,200 Btu/kWh would have a variable cost as follows:

- 7,200 Btu/kWh × $4.10/MMBtu × 1 MMBtu/1,000,000 Btu × 1,000 kWh/mWh = $29.52/mWh
- Variable operations and maintenance (O&M) would be $2.83/mWh (2010 dollars)
- Total natural gas plant variable cost = $32.35/mWh

The variable cost of power of $32.35/mWh is the addition of the fuel cost of $29.52/mWh, and the variable O&M cost of $2.83/mWh. The preceding calculation is a simple analysis that doesn't include the cost of any emission allowances that will be required. If the forward price for power were only $20/mWh, it would not make sense to dispatch the power plant, since it would not cover its variable costs. However, if the forward price for power were $50/mWh, it would make sense to dispatch the power plant, since it would make a $17.65/mWh ($50/mWh – $32.35/mWh) contribution to fixed O&M costs, return to debt, and return to equity.

As a comparison, a typical 500-mW coal plant would have a heat rate in the range of 10,000 Btu/kWh and would have a variable cost as follows:

- 10,000 Btu/kWh × $2.50/MMBtu × 1 MMBtu/1,000,000 Btu × 1,000 kWh/mWh = $25.00/mWh
- Variable O&M would be $5/mWh (2010 dollars)
- Total coal plant variable cost = $30.00/mWh

A power plant is actually a spread option since the selling price for power has to exceed the variable O&M and fuel cost. A power plant can be expressed as a spread option by the following equation:

Max [Selling price for power − (Fuel cost + Variable O&M), 0]

To forecast how frequently a natural gas power plant would operate using a spread option model, it is necessary to determine the individual volatility for both electric and natural gas. Volatility can be found by using historical prices, generalized autoregressive conditional heteroskedasticity (GARCH), principal components analysis, and calculated from at-the-money options. The correlation between electricity and natural gas can be found from historical data. A Monte Carlo simulation is then run to generate a series of random numbers, which are based on the volatility and correlation of electricity and natural gas previously calculated. These random numbers are then used to calculate future power and natural gas prices and power plant dispatch. The use of a single value for correlation between electricity and natural gas fails to take into account a number of factors that affect the price of power. This analysis can be further expanded by including or estimating weather, load, jump diffusion, mean reversion, and price floors. Each of these variables would result in a changing of the correlation assumptions and require additional Monte Carlo simulations.

Evaluating the future price and demand or volume risk for a power plant is a very challenging exercise. As previously mentioned, there have been a number of recent black swan events (a black swan event refers to an unpredictable occurrence) such as a large supply of shale gas reducing power prices, an unanticipated economic downturn, and the future potential of demand-side management (DSM) to reduce overall demand for power. Even though events of this type used to be thought of as a low-probability occurrence, they can wipe out the equity and some or all of the debt of a project and result in bankruptcy. These types of issues can cause models to produce results that could greatly overstate the future price of power. If a lender has sized the amount of debt to be provided to a particular project based on this type of electric power price forecast, the project could default unless the sponsor was willing to inject additional cash equity into the company. Having been burned in the past by this issue, lenders are concerned about lending large amounts of debt to power projects based on a power price forecasted by a model.

The variable cost of a power plant is equivalent to the strike price of an option. In most U.S. power markets, a gas-fired combined-cycle power plant sets the marginal price of power and as a result is an at-the-money option. Because a coal plant typically has a higher heat rate and a lower cost of fuel, it has a lower variable cost of power than a natural gas plant and is considered an in-the-money option. A simple-cycle power plant has a high cost of power due to its higher heat rate and is considered an out-of-the-money option. Volatility brings out-of-the-money options into the money. The volatility for power pricing is reduced when there is an oversupply of generating facilities and/or natural gas supply. In this scenario, a power plant will be forced to rely on its capacity payment. As a result, a simple-cycle power plant may experience more hours of operation than would be expected due to a spike in temperature or the forced outage of a nearby power plant. The simple analysis of calculating the strike price and comparing this to the forward curve is referred to as an *intrinsic valuation*. Since the future price for power is determined by a number of different variables, the actual price for power could be higher or lower. This type of analysis is referred to as an *extrinsic valuation*.

This extrinsic or real option approach to valuing a power plant is the same way that one thinks about a call option on a stock. If the stock of Microsoft is trading at $50 per share and I have a call option with a strike price of $100, this gives me the right but not the obligation to buy the stock at $100. Since I can purchase the stock at only $50, the call option is worthless. The value of the call option will depend on the length of its contract life. If the call option has a five-year life, there is a probability that the price of a share of Microsoft could exceed $100 per share and the call option would come into the money. Unlike call options, which are relatively inexpensive and very quick and easy to purchase, the cost to develop, permit, and finance a power plant is very expensive and time consuming. Since a power plant can have an operating life that exceeds 30 years, its call option value can be very large.

When evaluating a power plant, it is critical to review its air permit to make sure that it will be allowed to dispatch when it is called upon. This task is accomplished by reviewing all of the air emissions produced by the plant on a lbs/hour basis and converting to a tons/year basis by multiplying 24 hours/day by 365 days/year or 8,760 hours/year. The calculated tons/year for each pollutant is compared to the actual tons/year limit in the air permit. This exercise will show if a particular power could be constrained on its ability to operate due to its air permit. A gas-fired combined-cycle power plant might only be able to operate in simple-cycle mode or on oil for a limited number of hours. Since a number of regions in the United States are nonattainment (e.g., exceed federal standards) for certain air pollutants, older vintage air permits are very valuable. There is also a possibility that a power plant with an existing air permit would be given carbon dioxide (CO_2) allowances under a future cap-and-trade program as opposed to having to purchase them.

FINANCING A POWER PLANT

When determining how to finance a particular power plant, IPPs will consider a merchant sale approach as opposed to a full or partial sale of power to a third party. As in other industries, the forward sale of a power plant's electricity production at a fixed price for a fixed period of time is known as *hedging*. By fully contracting the entire output of a power plant, it is possible to achieve levels of debt of 80 percent or higher depending on the yearly debt coverage ratio. In a project financing, the debt coverage ratio is calculated by EBITDA/principal and interest payment. A power trader would feel that a developer had given up all of his upside by taking this approach. His thinking is that there will be future spikes in the price of power and the project will not be able to capture this spike since it has contracted all of its output forward with a third party. The trader is missing the point that debt is cheaper than equity and that as the project approaches a 100 percent debt capital structure, its internal rate of return is undefined. This type of fully hedged off-take structure also allows smaller developers without a balance sheet to finance their projects. Since renewable power projects generate less EBITDA than a fossil power plant, debt is at lower levels.

Another way to think about this capital structure is that power project developers were able to finance their projects with a large amount of debt since they had signed a long-term power purchase agreement (PPA) with an above-investment-grade-rated electric utility. This long-term PPA provided lenders with the credit support necessary to provide maximum amounts of nonrecourse debt. Utilities are concerned that these PPAs will be treated as imputed debt by the rating agencies and that they would be downgraded in the future. Utilities are not able to earn a return on rate base if they purchase power from an IPP. These PPAs are treated as a pass-through cost.

Renewable power projects create a large number of tax benefits, which can be sold to a tax investor under a partnership flip structure. These benefits can be looked as a source of equity capital funding for an entrepreneurial or corporate renewable power project developer without a tax appetite, provided courtesy of the U.S. Treasury. Part of the capital required for a renewable power project is funded by project finance debt, another part by tax equity, and the remainder by actual cash equity. A flip structure allows for a disproportionate allocation of the cash and tax benefits produced by a renewable power plant. Cash includes the EBITDA produced by the power plant, and tax benefits include depreciation, production tax credits, and debt interest deductions. Developers typically don't have a large tax appetite and, as a result, are allocated some of the cash benefits and a small number of the tax benefits until the time when the tax investor hits his hurdle rate and or most of the tax benefits are used up. At this point, a "flip" would occur and the cash and tax benefits would be reallocated between the developer and the tax investor.

The simplest way to determine the value of a power plant is by obtaining a fixed price swap for the power it produces. This results in giving away any upside in the price for power in order to get downside protection. Another approach is to purchase put options, which become more valuable as the price of electricity drops. The downside of this approach is the big up-front cost. A combination of puts and calls can also be sold and purchased in order to create a cashless collar. Finally, a series of put options with different strikes can be sold and purchased to create a put spread. Because the natural gas market is more liquid and has longer-dated pricing than the electricity market, forwards and options on natural gas are sometimes used to hedge power plants. A long-dated natural gas curve can be used to estimate the price for power long past the five-year time frame that is typically quoted. Basis risk, along with the correlation between natural gas and electricity, has to be considered in this case.

HEDGE PROVIDERS

Some IPPs will finance their projects based on a heat rate call option or tolling agreement. A tolling agreement requires a project to operate at the pleasure of the hedge provider. It can also be looked at as renting out the use of the power plant. Under some tolling agreement, the project receives a fixed payment whether or not it dispatches, and fuel and operating expenses are a pass-through cost to the offtaker. In this type of contract, it is important that the power plant provide an accurate estimate of its heat rate. If too low an estimate is provided, then the plant will not fully cover its fuel cost and it will be called to dispatch more than if it had provided an accurate heat rate. As long as an offtake contract is of sufficient length, credit quality lenders will provide high levels of nonrecourse debt to a project. If the hedge provider is an investment bank, the credit group of the bank may also participate in providing leverage to the project.

Individual power plants are frequently not located at liquid trading hubs or nodes. As a result, a power plant might have to incur the cost/basis risk to get power from a particular location to a node at which the hedge provider is willing to purchase the power. This can also be the issue for the supply of natural gas to the project. The project would have to pay the transportation cost to get natural gas to the power plant. Power plants also have a ramp rate and a requirement for starting natural gas and O&M as the plant comes up to full load. If the ramp rate is slower than competing power plants, this could reduce the overall liquidity, marketability, and optionality of the power plant. The hedge provider will also be interested in how much a particular power plant can turn down its heat rate and emission levels at different operating points before it has to be fully shut down. Start and operating costs can be reduced if the project can continue to be operated at low loads and is then able to quickly reach full load again rather than from a cold start. A power plant typically has a minimum shutdown time of six hours before it can be restarted. Hedges that are provided from nonregulated entities are typically a maximum of five years in length. This is due to the fact that the hedge provider itself has to hedge its positions, and power markets are not liquid beyond a five-year period.

In any sort of hedging arrangement, it is important to keep in mind the cost of hedging and the counterparty risk. The collateral demanded can be in the form of letters of credit or cash, and there can be obligations to supply power even if the power plant is not operating. As the price of power increases, it becomes more important that a particular power plant continue to operate. This requires larger and larger amounts of collateral to be posted. This could create a liquidity situation and result in a default for a highly levered power plant. Recently, power traders have been willing to take a first- or second-lien position pari passu to existing first- or second-lien debt providers. This reduces the amount of any cash collateral or letter of credit required and introduces intercreditor issues with other lien lenders. The challenge for a power trader is that he doesn't know what his exposure will be over the life of the hedge. However, a secured lender knows from the beginning his exposure to a particular power plant. Power traders determine their risk by calculating loan to value, extreme price scenarios, and a default scenario.

Hedge providers also consider a concept of right-way risk. This is based on the concept that as the price for power rises, the asset that produces that power will go up as well. However, an operating failure at a particular power plant could require an IPP to obtain replacement power at a high cost. In order to reduce right-way risks, hedge providers don't want a power plant to hedge all of its power output. If a power plant files for bankruptcy, the nondebtor hedge provider falls under a safe harbor and, as a result, is not governed by general bankruptcy rules and can immediately terminate the hedge contract. IPPs have remarked that in some cases it is so expensive to hedge a particular project that it might be better to leave a particular project unhedged. IPPs must also consider the optimum sizing for a particular project. If a project is too large, it could actually reduce system constraints to a point that the future price received for power would be lower than expected. Future diseconomies of scale could be determined by transmission constraints or fuel supply issues for a biomass power plant.

The price that can be paid for power to a project will depend on a number of issues. One issue is the ability to interconnect to a substation that serves more than one power market. Florida Power and Light (FPL) took this approach on its Sunoco refinery cogeneration project. The project's energy is sold to the PJM power market, while its capacity is sold to the New York Independent System Operator (NYISO). In this case, the NYISO places a higher value on capacity than PJM, and the project's location allows it to take advantage of this situation. As a result of the NYISO rules, Long Island Power Authority (LIPA, an electric utility that serves Long Island, New York) agreed to a long-term capacity sale, which allowed FPL to increase the amount of leverage in its capital structure. If the project is a renewable power project, its renewable energy credits might be more valuable in one market than another. Older projects that were developed before certain power markets were created might not have been able to take advantage of this opportunity in the past. If a project is at the end of its original PPA or is distressed, this is a way to increase its value.

OPPORTUNITIES WITH DISTRESSED RENEWABLES

Renewable power projects purchased at distressed prices have long-term viability due to future electric load growth, future carbon and other air emissions tax/cap-and-trade programs, and future increases in the demand for natural gas. Projects of this type have been under competitive pressure for the aforementioned reasons that should prove to be temporary factors. The investment opportunity during this period of reduced power demand and low natural gas prices is to purchase good-quality distressed renewable energy assets since the returns will eventually be justified when this cycle reverses.

Private equity funds typically can't use the tax benefits created by wind and solar power projects. This is due to the fact that their limited partners are not taxpayers. Most private equity energy-focused funds don't have the turnaround/operating skill set, distressed debt analysis, legal, and trading skill set to acquire distressed power projects. They are restricted by their fund documents to investing only in healthy power plants. Their limited partners would also feel that distressed renewable investments are a style drift and would ruin their correlations with existing alternative class and traditional investments. There is currently no private equity/control distressed fund that is solely focused on distressed renewable and natural gas–fired power plants. Mutual funds and some hedge funds have daily redemptions. This type of short-term lockup would not work for a renewable or fossil power investments.

The investment, regulatory, technical, tax, operating, and restructuring challenges involved with distressed renewables makes this type of investment challenging for generalist funds with limited lockup periods. The opportunities would be controlling ownership stakes in distressed wind, low-impact hydro, geothermal, biomass, solar, natural gas–fired power plants, and late-stage stressed development projects. The concept is to buy distressed assets at a substantial discount to par/replacement cost. These assets will then be restructured and refinanced. Investment opportunities would be sourced from the following:

- Purchasing distressed renewable power project finance loans from banks and insurance companies.
- Purchasing renewable power project finance loans from banks that go out of business, exit power project financing, or are forced by their regulator to sell their distressed loans.
- Purchase underperforming corporate orphans or corporate divestitures/carve-outs.
- European projects that become distressed as a result of a reduction of the feed-in tariff after commercial operations.
- Locating future investment opportunities early by correcting existing rating agency ratings downward based on extensive energy, credit, and restructuring experience.
- Acquiring late-stage development smaller renewable power projects with strong standard offer pricing.
- Acquiring late-stage development renewable power projects from underfunded and underresourced developers and private equity funds.
- Purchasing distressed natural gas midstream and storage opportunities.
- Purchasing power plant assets from ongoing Chapter 11 cases in a 363(b) sale.
- Providing debtor-in-possession financing in order to gain ultimate control of a power project.
- Acquiring and repowering stressed operating wind power projects in order to restart the production tax credit and five-year modified accelerated cost recovery system (MACRS) depreciation.
- Restructuring operating renewable power projects with uneconomic PPAs and capital structures either in or out of bankruptcy court.
- Acquiring the fulcrum debt security of stressed, distressed, and bankrupt natural gas–fired and renewable power plants in order to ultimately own the restructured company.

The first- and second-lien bank loans of natural gas–fired power plants are widely traded since they are usually over $300 million (par) in size. The debt for a typical renewable power project tends to be smaller in size and, as a result, is not widely traded. The entire debt issue for a renewable power project may also be held by an insurance company. It might be possible to have a debtholder of an existing renewable power project roll its existing debt into the new capital structure of a restructured entity. The fund objective will be to acquire the entire fulcrum security or enough of the fulcrum security to determine the pendency of the restructuring/Chapter 11 case. A blocking position is achieved by owning one third in amount of the fulcrum security or two thirds in amount to control. If it is not possible to obtain a sufficient amount to control the restructuring, then the fund could enjoy the return from an increase in the price of the debt it owns.

The control issue may be solved on wind power projects by the fact that their loans are typically controlled by a relatively small number of project finance banks and don't trade. The thesis would be to approach the agent bank on a distressed wind power project and convince the bank group to roll the debt into a new capital structure along with a possible new equity contribution. This situation could occur when a wind project had selected the production tax credit (PTC) option and overestimated wind production and underestimated maintenance, resulting in a large drop in EBITDA. This would require a new capital structure and technical fixes. It is not un-common on smaller renewable projects where the entire fulcrum security is controlled by one party.

It may be necessary to find third-party tax-based investors in order to monetize any PTCs, investment tax credits (ITCs), and five-year MACRS depreciation that are produced by renewable power projects. Most limited partners in private equity funds are not taxpayers and, as a result, can't use the tax benefits of a renewable power plant. If these investors are subject to the alternative minimum tax, this will limit the value that they can place on any tax benefits. Until the end of 2010, investors were allowed to convert the ITC to a cash grant. If this is not extended, investors without a tax appe-tite will have additional tax monetization challenges. The investor would leverage third-party distressed trading, power and energy trading, and proj-ect finance lending.

SUMMARY

Understanding the energy business means knowing how a power plant can be financed and hedged. Understanding the basics of fossil power plant eco-nomics for both regulated and independent power generators, revenue requirement calculations, hedging, and trading help investors evaluate whether a renewable power project can be developed and financed.

Chapter 3 further develops these concepts by reviewing tax issues for renewable power plants.

The Challenges of Renewable Power Projects

Buyers want to buy assets and sellers want to sell stock.
—Arthur Rosenbloom, senior consultant to CRA

TAX ISSUES

Tax issues are always an important part of any energy transaction. Sophisticated, smart investors like Carl Icahn and Warren Buffett have always used the tax code to their advantage. Private equity investors want an asset deal that allows them to write up the value of the entity being purchased. These investors should also use some leverage in order to further increase the tax shield. In this case, the tax code is used to enhance the overall deal economics as opposed to being the reason for doing the deal. Renewable power projects are essentially tax-driven deals that depend on either the monetization or use of tax attributes. Unfortunately, using these attributes can be a challenge.

Wind projects produce a small amount of earnings before interest, taxes, depreciation, and amortization (EBITDA); a large amount of depreciation; and either an annual production tax credit or an up-front investment tax credit. This reduces the amount of leverage that a lender can provide to wind projects and results in a large number of tax benefits that can't be used by most project developers. Renewable projects depend on individual states for renewable energy credits and the federal government for tax credits and accelerated depreciation. As of the end of 2011, the ability to convert the existing investment tax credit to a 1603 cash grant is currently scheduled to expire. Due to the current budget deficit and upcoming presidential election, it is highly unlikely that this program will be extended. This cash grant is equal to 30 percent of a project's investment cost.

Passive Loss

Most developers of wind projects are not able to use all of the tax attributes produced by a renewable power plant. As a result, third-party tax investors are required. Unlike debt financing, it is not just the cost of tax equity since most developers can't use the tax benefits that a renewable project produces. Debt investors will require a higher yield for their debt if they are located at a holding company and the tax equity is located at an operating company. Lenders will view this structural subordination to the tax equity as an additional risk that would cause a lower recovery in a bankruptcy. As a result, debt financing for the project will be more expensive.

If the underlying price for power is high enough, a developer can use the tax code as his equity and can obtain project finance debt. The challenge for renewable power projects is that they don't create enough EBITDA to be able to take a large amount of debt. Before the financial crisis, renewable developers often partnered with large companies that could use the tax attributes produced by renewable power plants. This type of investor is referred to as a tax equity investor. The cost of capital for a renewable power project is made up of the cost of project finance debt, tax equity, and investor equity. As a result of a reduction of tax equity investors, the required discount rate for tax equity investors has been decreasing. This has resulted in an overall increase in the cost of capital for renewable power projects. With the extension of the cash grant, this issue has been put off into the future.

Most wealthy and even corporate investors don't have the required tax appetite and/or corporate structure to use these benefits effectively. By way of example, a 50-mW wind project that had elected the production tax credit (PTC) would produce $2,759,400 in tax credits in the first year of operation. This is based on 50-mW $\times$ \$21/mWh $\times$ 8,760 hrs/year $\times$ 30 percent availability. As a result, an individual or corporation would have to have a yearly minimum taxable income of $2,759,400. The limited partners in private equity funds are not taxpayers, and they are not able to use either the investment tax credit (ITC) or the PTC. Private equity investors have been pushing for the cash grant program to be extended in order to avoid this issue.

It is only widely held (as defined by the IRS) C corporations that can use passive income. Internal Revenue Code (IRC) Section 469 limits the deduction of passive activity loss (PAL) by individuals, closely held C corporations, and other taxpayers (but not by widely held C corporations). Unlike individuals and other taxpayers, a closely held C corporation is allowed to offset PALs against active income from the same tax year. Portfolio income is not treated as active income or as passive income. The ruling allows the offset of different categories of income and losses (portfolio income and active losses) from different years, even though the taxpayer was closely held in some of those years. Under IRC Section 469, a closely held C corporation is a C corporation that, at any time during the last half of the tax year, has more than 50 percent in value of its stock owned, directly or indirectly, by not more than five individuals.

A large number of investments in renewable power projects can place a C corporation into an alternative minimum tax (AMT) position. AMT rules limit the number of deductions that a corporation can take. In this situation, the tax benefits that a renewable power project produces can't be fully used by a C corporation. A corporation that is not in AMT can afford to pay a higher price for the tax benefits produced by a renewable power project. AMT issues are also temporarily solved at the time of this writing by the cash grant for the ITC. Bankers have been known to say that in order to preserve their valuable tax attributes, they are careful which renewable power deals they invest in.

SPECIAL EXEMPTIONS

Various investors/developers have tried to work around the passive loss issue, at-risk restrictions, and publicly traded partnership rules in order to target retail investors. Unlike wind and solar, the oil and gas and low-income housing sectors have special exemptions.

Master Limited Partnerships

Oil and gas developers are allowed to set up publicly traded limited partnerships called master limited partnerships (MLPs). These MLPs are restricted on the amount of nonqualifying income they can earn. Only a certain amount of earnings in an MLP can be from electric power plants. An MLP is a very tax efficient structure, and most companies would become an MLP if they could. Unlike renewable power projects, oil and gas projects generate a much larger amount of EBITDA, which allows the project itself to utilize the tax benefits.

In the current market, renewable power projects have been electing the ITC as a cash grant and carrying forward the five-year modified accelerated cost recovery system (MACRS) depreciation. Since renewable power projects generate only a small amount of cash, they are not able to use the depreciation schedule in the current and have to carry it forward. This creates a net operating loss (NOL). Developers have found that it is not worth the cost or the time to bring in an investor to purchase this NOL. If a developer files a consolidated tax return, he might be able to use this NOL on the cash generated from other projects in his portfolio.

Wind developers without tax appetite have been selecting the cash grant and carrying forward the NOLs from the five-year MACRS depreciation and debt interest deductions to future years. Assuming the cash grant is not extended, wind developers may have to bring in third-party investors as general partners in order for these firms to be able to fully use the tax benefits. These investors will have to get comfortable with the risk of the wind project and, unlike limited partners, theoretically are exposed to all of the obligations of a particular wind project. If some sort of project disaster occurs, this general partner could be liable for damage awards. Both limited and general partner investors need to have a very large tax appetite.

Leveraged Lease Structure

Another approach is to use a leveraged lease structure. A leveraged lease involves a debt provider such as an insurance company or traditional bank lender and an equity investor with a tax appetite. It allows for the use of all of the tax attributes and can provide 100 percent financing. This approach is now possible for wind power projects since an ITC election is now possible, as opposed to a PTC. PTCs can't pass through a leveraged lease. In the past, only solar power projects could consider using a leveraged lease structure because they qualified for the ITC.

In a typical leveraged lease, the developer sells the project asset to a trust that is held by the owner trustee. This trust is established for the benefit of the lessor or owner trustee. The owner trustee is usually a financial institutional such as GE, JPMorgan, or even a local bank. It can also be a corporation with a large tax appetite. The owner trustee then leases the project assets back to the developer. The developer grants a security interest/mortgage in all of its rights relating to the project (including the transmission lines, contract rights, and revenues under the project contracts) to the owner trustee to secure the obligations of the developer under the lease. The owner trustee then issues notes or borrows debt and grants a back-to-back security interest in the project assets to the loan trustee for the benefit of the debt provider. In some older power purchase agreements (PPAs), the purchasing utility may have been granted a second lien on the project's assets to secure the obligations of the developer to the utility under the PPA.

If the project becomes distressed, the trust would file for bankruptcy. The cooperation of the owner trustee would be required since it would be the debtor in this case. Since most renewable projects are relatively small in size, the approach would be to file a prearranged bankruptcy. The sponsor of the plan would agree to redeem and refinance any outstanding debt, depending on the agreed-upon valuation. The indenture and lease would be terminated. As part of the bankruptcy plan, all of the contracts to which the owner trustee is a party that are no longer desired would be terminated and any liens on the project assets would be wiped out. The utility could object to this if it felt that the enterprise value of the company broke past the lien in front of its claim against the estate.

The owner trustee through the trust is the tax owner. In the prior bankruptcy situation, the owner trustee would be treated as selling its ownership interest for the unpaid balance of the outstanding debt. This would result in a taxable gain to the owner trustee equal to the amount of debt less its tax basis in the project asset it owns. The calculation would be the amount of debt multiplied by an assumed combined federal, state, and local rate of 40 percent. This assumes that the owner trustee could not offset such income in whole or in part with NOLs and/or capital losses. This would not be an absolute cost to the owner trustee because it would have had to pay that tax over the remainder of the debt term if no default had occurred. This situation results in phantom income for the owner trustee. The tax disadvantage to the owner trustee would be the present value of accelerating the phantom income on transfer of its interest.

The challenge with leveraged leases is that they are expensive, complicated, and difficult to arrange. If a leveraged lease project becomes distressed, it also requires an extensive effort to restructure. The lessor can also be exposed to depreciation recapture if it is terminated early. This complexity can make for an interesting distressed investment opportunity. In most bankruptcy cases, the lessor would be the debtor and would control the pendency of the case.

"Partnership Flip" Transactions

The downside of doing a sale-leaseback versus a partnership flip is that it costs more for the developer to get the project back. After the lease ends, the developer can continue using the project only by purchasing it from the investor. There is also a different risk allocation in leases compared to partnership flips. The developer may be required to give the investor a broader indemnity against loss of tax benefits in a lease.

Numerous wind projects have used a partnership structure to allocate tax benefits. "Partnership flip" transactions are a way of bartering the tax subsidies to an institutional equity investor, who can use them in exchange for capital to build their projects. A partnership structure allows for a disproportionate split of the cash and income accounts between the developer and the tax equity investor. It is hard to do a partnership flip transaction with just depreciation. The depreciation can be carried forward for up to 20 years and used when the developer has income against which to offset it. Alternatively, a developer might enter into a tax equity transaction to try to convert the depreciation into cash.

Each partner has a "capital account" and "outside basis" that are limits on its ability to absorb tax benefits. Hence, the cash grant has tended to be thrown into the partnership also. The tax equity investor essentially bridges the cash grant. Partnership structures are broken down further into pretax after-tax partnership structures (PAPS)—partnership flip transactions in which the tax equity investor pays the full purchase price to buy into the deal up front—as opposed to over time in a "pay-go" structure (G-PAPS—a PAPS deal with a cash grant).

A typical partnership structure would have the developer contribute 1 percent of the equity and the strategic tax equity investor contribute 99 percent of the equity. If the project elected the PTC, they would also be split 1 percent to the developer and 99 percent to the tax equity investor. Distributable cash would be initially split 1 percent to the developer and 95 percent to the strategic tax equity until its hurdle rate was hit. At this point, the distribution would flip to 95 percent to the developer and 5 percent to the strategic tax equity investor. Tax benefits/liabilities would also be initially split 1 percent to the developer and 99 percent to the tax equity investor until its hurdle rate was hit. At this point, the distribution would also flip to 95 percent to the developer and 5 percent to the strategic tax equity investor.

New Market Tax Credits

Developers have also been considering the use of new market tax credits (NMTCs). NMTCs are intended to promote private investment in low-income neighborhoods by providing an investment tax credit equal to 39 percent spread over a period of seven years. This 39 percent is broken down into 5 percent in each of the first three years and 6 percent in each of the last four. There is a minimum investment period for NMTCs of seven years.

SUMMARY

Renewable projects produce only a small amount of EBITDA—they are dependent on leveraging the tax code. The economics of renewable power projects are based largely on the tax credits they produce. For an entrepreneur, they are her equity; for a corporation with a tax appetite, a key part of its return from investing in renewable power projects. Investing competitively in renewable power projects is difficult for C corporations in AMT or with no tax appetite. Markets based on tax credits and investor tax appetite can have boom-bust cycles, which can provide interesting distressed investment opportunities.

In Chapter 4, we discuss assessing the risk of power projects.

Risk Assessment for Power Projects

We need more civil engineers, less financial engineers.
— Larry Summers, director of the White House
National Economic Council for President Obama

Both fossil and renewable power projects are very complex and capital intensive and require a large amount of time to analyze and understand. In addition, renewable projects involve the added complexity of numerous tax attributes and the frequent need for outside tax equity investors.

Unlike some other investments, they require numerous permits and approvals, which can take years and a large amount of risk capital to obtain. Project development internal and external expenses can increase very quickly and must be constantly managed. Internal expenses involve the labor and benefit cost for the developer's or project sponsor's own employees, and external costs include attorneys, lender's engineers, owner's engineer, environmental engineers, and market consultants.

Investors in power projects have to consider both their own financial cost of capital and opportunity cost when they review potential deals. This is based on the fact that they could be developing or diligencing other projects when they are focusing on a particular deal. A simple "make versus buy" analysis can often help investors evaluate potential investments and potential returns.

Whether a power project is based on fossil or renewable technology and how it ultimately will be financed, it is critical to develop a project-specific risk assessment and mitigation analysis. An analysis of this type would be created in addition to a financial model and a development budget and schedule. Development dollars are risky, expensive dollars, and considering these issues during initial due diligence can save an investor a lot of future heartache.

PROJECT RISK ASSESSMENT AND RISK MITIGATIONS

The following risk assessment shows the key risks and mitigants for the restructuring of an operating distressed hydropower plant. The hydro project has a long-term power purchase agreement (PPA) that no longer supports its outstanding debt. The PPA is currently priced at a large discount to avoided cost. The project is located at an electric substation that potentially allows it to sell power to two separate independent system operators (ISOs). The existing PPA can be rejected in a prepackaged Chapter 11 reorganization, and the sale of power could be recontracted with another entity.

Since the hydro project is relatively small, it is important to use a prepackaged as opposed to a "free fall" Chapter 11. It is hard for smaller firms to bear the cost of a traditional Chapter 11 proceeding. The overall hydro financial structure is complex since it involves a first lien loan, leveraged lease, and a second lien, which was granted to the power purchaser. Table 4.1 shows an analysis developed by the project sponsor's management team to help think through the key issues. The analysis includes both technical and financial issues.

PRECOMPLETION RISKS/MITIGANTS

Before a power project is financed or even starts operation it has to complete a number of key tasks. One of these tasks is the acquisition of all air, water and local approvals. This expense is typically carried by the developer and can easily run into millions of dollars just before financial close. In order to achieve some form of nonrecourse financing, it is necessary for all of these permits to be out of any appeal period. This is due to the fact that lenders will consider permitting to be equity and not debt risk.

Potential project risks, and proposed mitigants, are further discussed in Table 4.2. Although developers and project sponsors will seek to develop the project in a conservative, comprehensive manner with attention to details, every economic undertaking involves risks that may not be quantifiable or discernable at specific points in time.

TABLE 4.1 Analysis of Key Issues

Risk	Primary Risk Bearer	Mitigants
Market/Offtake factors	Project company	The existing PPA with the electric utility will be rejected by foreclosing on the first lien and a possible prepackaged bankruptcy. Due to the project's unique power transmission location, power can be sold to two different ISOs.
Future project operations and restructuring staffing	Project company, O&M company	There is an existing plant manager at the site who has worked on the project since the start of operations. A third-party operator can also be hired.
Project operating permit and low-impact hydro certification	Project company, lenders	The project Federal Energy Regulatory Commission license doesn't expire until 6/28/2037 and can be relicensed after this date. We have received a preliminary opinion from the Low Impact Hydro Institute that the project would qualify as a low-impact hydro project. This would be one of the conditions required to potentially obtain Class 1 renewable energy credits.
Electric utility may contest that the project's enterprise value breaks at the first lien	Project company	We could negotiate with the electric utility to resolve this issue through minimal payment or, if unsuccessful collect on the first lien loan in a bankruptcy filing.
Project mWh production	Project company, lenders	Downside production sensitivity of 140,000 mWh, which is c. 17 percent less than the lowest annual production over the past 10 years, results in an unlevered internal rate of return of 8.2 percent under our future natural gas price assumption. Project has achieved a median net cash flow of approximately 65 percent over the past six years. Capacity will be bid conservatively and energy on an as-available basis.

TABLE 4.2 Possible Construction Period Risks

Risk	Primary Risk Bearer	Mitigants
Approvals and permits	Project company; engineering, procurement, and construction (EPC) contractor (for certain permits only)	The project company will obtain all major central government and provincial approvals with all appeal periods expired prior to financial close and funding. The project will be designed to exceed World Bank standards.
Cost overruns	EPC contractor, project company	Fixed price turnkey construction agreement, the project company will have approval rights over all change orders with, under certain circumstances, the consent of the lenders; a construction contingency and likely sponsor cost overrun support
Technology	EPC contractor	Only proven technology will be employed; the turnkey construction agreement will include certain warranties; equipment suppliers will provide certain warranties.
Construction delays	EPC contractor	Turnkey construction agreement contains liquidated damages for delay in start-up, incentives for early completion, insurance for force majeure events, construction contingency, and likely sponsor cost overrun support.
Transmission line delays	EPC contractor	Date certain obligation per the turnkey construction agreement, minimal upgrade required, project site all within the refinery site; construction contingency and likely sponsor cost overrun support.
Project performance (per performance test)	EPC contractor	Turnkey construction agreement specifications, turnkey construction agreement liquidated damages for efficiency and output shortfalls; bonus incentives for project performance in excess of turnkey construction agreement specifications.

Land use rights	The project company will negotiate a site lease with refinery prior to financial closing. All of the real property required is within the control of refinery.
Interest rates.	The project's final capital cost, upon which the electricity tariff will be based, will reflect the actual interest during construction (IDC), fixed rate loans, as available, will be utilized; interest rate hedges, if appropriate and available may be utilized; the construction contingency and sponsor cost overrun support can be utilized to cover higher-than-projected IDC.
Insurance-related claims	Comprehensive all-risk insurance program will be put in prior to financial close.
Penalty payments and termination under the PPA	Relief will likely be allowed under the PPA for force majeure events and governmental action or inaction. The PPA will likely allow the project company to pay penalties in lieu of termination. Date certain completion turnkey construction agreement.
Force majeure	Force majeure declarations will be provided for in the project contracts; delay and all-risk insurance will be purchased; the construction contingency and sponsor cost overrun support can be utilized to cover the cost of force majeure events.

Project company

Project company

Project company, EPC
contractor

EPC contractor, project
company, lenders

Power purchasers, project
company and lenders

TABLE 4.3 Possible Operational Risk Phase Risks

Risk	Primary Risk Bearer	Mitigants
Equipment defects	EPC contractor, project company	Turnkey construction agreement warranties; equipment vendor warranties.
Project availability	Project company, O&M company	Experienced O&M company; independent engineer opines on projected availability; O&M company bonus/penalty based on performance; proven technology with operating history; fuel supply contract bonus/penalty structure.
Plant efficiency (heat rate)	Project company, O&M company	Experienced O&M company; independent engineer opines on heat rate projections; O&M company bonus/penalty based on performance; proven technology with operating history.
Operations and maintenance costs	Project company, O&M company	Experienced O&M company; proven technology with operating history provides well founded basis for estimated O&M costs; O&M company bonus/penalty based on performance; maintenance programs and yearly operating budgets.
Regulatory risk (power sector)	Project company, lenders	Long-term take or pay contract negotiated directly with power purchaser; power purchaser payment obligations are absolute and unconditional; existing third-party contracts are likely to be respected in a deregulated power sector; deregulation not likely to change status of power purchaser within its service territory; the refinery PPA will be unaffected by deregulation; project company's power price will be competitive with other resource options.
Fuel price	Refinery as fuel supplier, the power purchase agreement allows for fuel cost pass-through	Pricing formula will be agreed to within the project economic constraints; up-front prepayment for the fuel is a possibility.

Fuel supply	Refinery as fuel supplier	Long-term fuel supply contract; project has first priority to petroleum coke; at least a 30-day stockpile reserve; an independent engineer will opine on the projected annual petroleum coke production.
Electricity demand (competitive power rates)	Power offtakers	Long-term take or pay contracts with power purchaser and refinery, proven demand by both offtakers; power market study will verify that the power price will be competitive in the future.
Credit status of power purchaser and refinery	Project company	Both power purchaser and refinery are established, respected and creditworthy corporations; credit standing acknowledged by the financial community.
Transmission interruptions	Power purchaser, project company	Power deliveries to refinery do not require transmission; power purchaser to take delivery of electricity at the high side of the step-up transformer on the project site; open access transmission in the Philippines.
Inflation	Power offtakers	PPAs provide for operating and energy payments to adjust annually with both Philippine and U.S. inflation as appropriate.
Foreign exchange rates	Power offtakers	Payments per the PPAs to cover offshore obligations will be paid in pesos but denominated in U.S. dollars on the payment date.
Foreign exchange availability and transferability	Project company	Central bank approvals to convert pesos to dollars will be obtained. However, delays in conversion may occur. Repatriation of investment returns on the part of foreign investors is permitted under Philippine law.
Interest rates	Power offtakers	Interest costs are covered in the power purchaser PPA by the capacity and fixed operating payments; interest rates will be fixed, where possible; interest rate hedges will be utilized, if available and appropriate.
Availability of insurance	Project company	Insurance adviser will opine regarding the project's insurance requirements prior to financial close.

(continued)

37

TABLE 4.3 (*Continued*)

Risk	Primary Risk Bearer	Mitigants
Environmental matters	Project company, power offtakers	Project designed to exceed both World Bank and Philippine standards; PPA pricing is adjusted for changes in law; project design will be flexible enough to allow for installation of additional emission control equipment if required; refinery to provide environmental indemnity for preexisting site conditions.
Change in law	Power offtakers	PPA structures provide for tariff adjustments to reflect increased costs arising from changes in law.
Force majeure	Power offtakers, project company	Under the PPA, a force majeure event will not excuse power purchaser and refinery from making firm payments as reduced by any insurance proceeds received. Insurance coverage will also be in place for natural disaster force majeure events.
Power purchaser franchise expiration	Power purchaser, project company	Power purchaser expects the franchise will be renewed, if not renewed, Power purchaser remains obligated to make firm payments under the PPA. The capital markets accepted this risk in the Quezon project. Project would still be able to provide electricity at a competitive price to a new purchaser.
Integration with the refinery; dependence on refinery	Refinery, project company	The obligation of refinery under the PPA, steam and fuel supply agreements are absolute, unconditional, and long term; the integration of the project with the refinery will enhance the competitive position of the refinery; refinery will take an equity position in the project; project revenue risk is diversified via the PPA.
Annulment of oil industry deregulation law	Refinery, project company	The obligations of the refinery under the PPA, steam and fuel supply agreements are absolute, unconditional, and long term; it is anticipated that the oil deregulation law will be reinstated with modifications.

Table 4.2 summarizes the possible risks perceived by the project company to be developed in Southeast Asia during the construction period. This project involves the development, permitting, financing, and construction of both a circulating fluidized bed (CFB) boiler and the upgrade of an existing refinery. The refinery upgrade will also produce the fuel supply for the CFB. The CFB boiler will also sell steam and some of its electric power to the refinery. The remaining power will be sold to the local electric utility under a long-term PPA.

POSTCOMPLETION RISKS/MITIGANTS

After construction of the CFB project is completed, it enters the operations phase. Investors will typically evaluate the economics of a project over a 20-year period. The project can have an operating life that can greatly exceed this time period. The key concerns at this point include project performance under existing contracts, availability, capacity factor, and overall plant performance. These metrics help determine the project's ability to generate cash, which allows it to pay back its debt and equity investors. The fuel supply and its future pricing for a fossil plant or the future wind production forecast for a wind plant are critical. For example, a wind project can default on its loan if it overstates future wind production and understates maintenance cost. Table 4.3 summarizes possible risks to the project during its operational phase.

SUMMARY

For either fossil or renewable power projects, understanding and managing risk during development, construction, and operation is critical. The risk assessments in this chapter (overall, precompletion, and postcompletion) show how risk is reviewed and managed from initial due diligence all the way through the life of a project. It is critical to stop due diligence or development on unfeasible projects as quickly as possible in order to reduce both actual and opportunity costs. This type of analysis is required for both renewable and fossil power plants, and these issues and techniques will continue to be referred to. Recognizing risk, risk bearer, and risk mitigants are important steps in the multistep process of investing in both proposed and existing fossil and renewable power assets.

Chapter 5 discusses opportunities with municipal bond–financed power projects.

Exploiting Profitability of Distressed and Abandoned Municipal Power Plants

"All good deals are behind us" is a myth.
—Scott Unger, founder and managing director, EnerTech Capital

Because of the past three years of deep U.S. and European economic business recessions, huge job losses, and financial declines, as well as lack of construction and business sales, the U.S. Federal Reserve Board, which sets key interest rates, recently recommitted itself publicly to the entire nation, stating that it would maintain America's currently very low interest rates on corporate loans, consumer loans, and investment loans for at least the next year and probably longer.

These current low rates of interest due to the continuing U.S. recession provide significant opportunities for investors across the entire United States to take advantage of distressed municipal power plants at potentially bargain rates. The U.S. Federal Reserve Board's desire to keep interest rates low to enable the U.S. jobs market and housing markets to recover from their deep decline has led many financial funds and investors to be able to salvage troubled or previously abandoned or previously nonfinanced municipal waste-to-energy power plants at profitable rates.

In addition, new favorable environmental legislation and energy legislation at the federal, state, and municipal levels has made it possible to finance or refinance municipal waste-to-energy projects in ways that were not possible before. This is due to the fact that "waste-fuel power plants" can qualify for either municipal bond debt or the investment tax credit (ITC)/accelerated tax depreciation. Because municipal bond interest rates are too low, the ITC/accelerated tax depreciation is selected by most investors or funds today since it is currently a much more valuable financial benefit to them.

WASTE-FUEL PROJECTS HAVE KEY FINANCIAL ADVANTAGES FOR INVESTORS

Waste-fuel power plants can be either municipally or privately owned. The state, city, county, region, public authority, public-private consortium, or a purely private company can be the initial instigator of the proposal for the waste-fuel power plant or the ultimate owner of the plant. Because a waste-to-energy power plant can be large and expensive, it must be funded to a very large extent by either municipal bonds or a mixture of private and public bonds. It can be backed partly by corporate guarantees, government guarantees, government grants, state, federal funds, or a consortium of multiple municipalities, government, or municipal loans in addition to private investors or corporations. Traditionally, a public power authority, whether state, city, regional, or federal in scope, was the primary developer of power plants in many regions of the United States. Recently, there has been a wider spectrum of merchant-owned or -funded waste-to-energy power plants and a rapid rise worldwide in the use of waste wood, waste coal, or other waste product leftovers from factories.

In this chapter, we explain specific roles and responsibilities of the different participants involved in planning, financing, designing, constructing, and managing a waste-to-energy power plant over its 30-year life.

DUTIES OF PROFESSIONALS IN A MUNICIPAL POWER PLANT

The key developer (or investor) owner of a private power plant usually has had extensive experience on various types of power projects or for a municipal power authority. Alternatively, he may have worked as an assistant to a major developer on one or more previous projects. The reason that this previous professional experience is so important in a private cogeneration project is that there are so many layers of regulation involved in planning, constructing, and completion of a cogeneration power plant that many entrepreneurs with no prior experience would be unable to navigate the maze of legal and regulatory permissions and deadlines, time schedules, and approval processes to be able to secure funding for the entire project in advance, without having usually an extensive track record of experience that banks, funds, and government approval agencies will expect and look for.

Second, municipal bond–backed power plants take two to five years to plan, permit, finance, construct, and manage. This is a long time for most investors to learn on the job for the first time. Therefore, other developers or consortia can usually operate on a fast track to capture that start-up investor's cogeneration project out from under them. That is why it is common to see teams formed to develop not simply one project on its own, but two to four projects at a time, so that while delays or road blocks are occurring in one project, certain members of the development team, or the engineering team or the financing team, can keep busy on parts of another power project so the delays are not simply dead time or totally wasted.

The investment banker or underwriter is one of the most active members of the waste-to-energy plant planning team. That is because they usually have raised multiple financings for a variety of projects in the past, along with other members of their firm. Although a waste-to-energy plant is a sophisticated engineering project with all kinds of professional detailed experts involved, ultimately, it is the ability to actually finance the project in a timely fashion that separates the winners from the losers. Many municipal bond–backed power plants can cost at least $200 million. They may easily hit cost overruns for a whole variety of reasons of construction delay or legal, regulatory delays. This can extend the total construction time of the project and can derail the project altogether. So many moving pieces in these complex and expensive projects must be able to be timed to come together on a tight schedule, the coalition of partners and financiers may not be able to hold together long enough to succeed in finally completing this original complex project. Some players drop out of the deal, and finding replacements in bad economic environments can prove excruciatingly difficult. That is why formerly troubled or abandoned municipal power plant projects are available today for financing.

Energy power plant investment bank underwriting teams often are not only experienced in working with the variety of specific states, energy departments, environmental protection departments, municipalities, finance directors, treasurers, municipal bond legal counsel, underwriter's counsel, issuer's counsel, and various municipalities' financial advisers, but also with a range of bank trust departments, who will usually hold the majority of the proceeds of the bond issue until monies are needed for the actual construction of the waste-to-energy power plant. Because brand new waste-to-energy power plants are often long term to plan, to finance, and to construct and are considered complicated to finance, they have often been sold via a negotiated deal between the bond underwriter and the lead manager of a specific investment bank, known to be specialized in complex bond financings, instead of being bid-out at a competitive auction, the way plain vanilla municipal bonds for general obligation tax-backed bonds and notes would usually be sold. Private negotiated agreement means that often the profit or "spread" earned by the investment bank on these more complex deals is greater than on public-auctioned municipal bonds.

In the past, in many bond underwritings, the lead syndicate manager of the group of investment banks participating in managing the bond sale would seek the extra job of "bidding out the bond escrow" to supposedly make sure that the best price was obtained by the bond issuer or municipality from the banks for investing and managing those bond fund proceeds for years until the funds were needed for actually constructing the power plant. Since then, ethical and legal problems were found in systemic collusions between various investment bank lead underwriters, financial advisers, and the commercial banks' insurance companies or financial firms about who would win the right to hold the bond fund escrows by ensuring extra profits to the winning escrow agent instead of maximizing the earnings on the escrow directly for the municipality. Industry-wide lawsuits and new regulation have settled this problem.

THE PROFESSIONAL FEASIBILITY STUDY ENGINEER

There is no universal format for independent engineering feasibility studies because, typically, feasibility studies are adapted to the particular type of facility that is being built (for hospitals, housing, public power, etc.).[1] Factors that impact study design include legal and regulatory requirements, geographic location, source of fuel or waste, regional demographics, or the total number and distribution of the total number of people in the geographic region. There are also important financial budgetary constraints on how much the community can pay for the power plant versus lowest feasible total dollar cost to plan, build, test, and maintain the required size and necessary amount of electric power needed to be produced by the power plant each month and each year in order to pay off the total bond debt borrowed to build and sustain the power plant for its useful life. It is typical in the waste-to-energy electric power industry for a private developer or municipality to discuss the particular focus and scope of the independent engineering feasibility study with the engineering firm employed to conduct the study.

---

[1] Robert Lamb and Stephen Rappaport, *Municipal Bonds Book* (New York: McGraw-Hill, 1986), 86–87, 156, 251.

The independent engineer conducts a sensitivity analysis to evaluate the ability of the project to provide adequate debt service coverage in the event of a reduction in spot market tipping fees or other increase in fuel cost. The results of this analysis are disclosed in the bond offering materials. In addition, investment banker practice is to provide potential investors in the bonds with CD-ROMs containing the spreadsheet data underlying the sensitivity analyses so that prospective bond purchasers can test any assumptions made in the bond offering materials that they wish.

Each time a nationally famous, long-term leading engineering firm performs an "engineering feasibility study," its professional reputation is on the line. The goal is to follow appropriate procedures for the conduct of an independent engineering feasibility study that is "inconsistent with recklessness." "Recklessness" is a breach of "The Legal Standard requiring a high level of High Duty of Care" that financial and engineering professionals must meet in the quality of their work.[2]

The investment bank reviews the independent engineering feasibility study to have a reasonable basis to believe that the project is viable and the representations in the bond offering materials are accurate. The investment bank must act in accordance with industry standards in making considered judgments with respect to these issues. There can be nothing reckless about the investment bank's consideration of and reviews with respect to the work of the independent engineer, the partnership, and others on the project.

DISCLOSURES OF RISKS IN THE BOND OFFERING MATERIALS

In accordance with industry standards, the bond offering documents typically provide to prospective investors the material risks involved in investing in the bonds. For example, the following is an example of a typical disclaimer on the cover of the official statement for a waste-to-energy bond offering:

> *PURCHASE OF THE BONDS INVOLVES A SIGNIFICANT DEGREE OF RISK. INVESTORS SHOULD READ THIS ENTIRE LIMITED OFFERING MEMORANDUM [LOM] TO OBTAIN INFORMATION ESSENTIAL FOR MAKING AN INFORMED INVESTMENT DECISION. [See "Risk Factors" MEMORANDUM TO OBTAIN INFORMATION ESSENTIAL TO MAKING AN INFORMED INVESTMENT DECISION.*

---

[2] Lamb and Rappaport, *Municipal Bonds Book*, 225–237.

*SEE "RISK FACTORS" BEGINNING ON PAGE ___ FOR A
DISCUSSION OF CERTAIN FACTORS THAT SHOULD BE
CONSIDERED IN EVALUATING AN INVESTMENT IN THE
BONDS.]*

In accordance with industry standards, the qualified sophisticated institutional investors who purchased the bonds are expected to conduct their own evaluation of the bonds, and particularly the risks. The offering documents for the bonds were more extensive than is the norm for municipal bonds with respect to risk disclosure, and they spelled out a large number of specific risks. Some offering statements for general obligation municipal bonds backed by taxes have minimal risk disclosure. Other offering statements for municipal revenue bonds often have a limited list of standard risk disclosures. A typical Limited Offering Memorandum [LOM] for a municipal bond offering lists over 20 pages of "risk factors." This review would have made it clear that the bonds were "unenhanced," not rated, not insured, not guaranteed, nor backed by bank letters of credit or state, city, or municipal taxes. They were, therefore, significantly riskier than municipal securities with a federal, state, or municipal or an insurance company's guaranty or letter of credit features or with total backup credit support.

The many categories of risks disclosed in the bond offering documents included risks relating to "Project Risk; Limited Recourse," "Limited Assets," "Construction Risk," "Solid Waste Market Conditions and Non-Authority Waste Requirements," "Governmental Regulation and Approvals; Loss of Permits; Environmental Matters," "Operating Risk," "Reliance on the Authority and the Member Cities," "Constitutionality of Waste Supply Agreements and Related Agreements," "Reliance on Power Purchase Agreement," "Qualifying Facility Status," "Third-Party Contract Risk," "Risks of Costs of Operation Rising Faster than Revenues," "Risks Relating to State Grant," "Reliance on Projections and Underlying Assumptions," "Factors Limiting Enforcement of Rights; Realization of Collateral and Enforcement of Judgments," "Cross-Default from Service Agreement to Project Site Lease," "Metal Recovery Revenues," "Adequacy of Insurance," "Loss of Federal Tax Exemption," "Additional Bonds; Financing Risks," "Potential Change of Control of Partnership," "Computer Modifications," "Absence of Market for Bonds," and "No Credit Rating."

Qualified sophisticated investors are willing to incur the levels of risk associated with and purchase "unenhanced" municipal bonds because of the significantly higher yields that the private energy companies or corporate partnerships who develop these projects will offer. Here is the standard disclaimer:

Although they may offer higher current yields than do higher-rated securities, low-rated and unrated securities generally involve greater volatility of price and risk of principal and income, including the possibility of default by, or bankruptcy of, the issuers of the securities.

Municipal securities such as these bonds are marketed and sold only to "qualified sophisticated institutional investors" because such investors have the experience and resources to understand, evaluate, and manage the significant risks that are involved. The bonds can be sold in a "private placement" to qualified sophisticated institutional investors who qualified as "Approved Institutional Investor[s]" with "at least $100 million in securities," "Qualified Institutional Investors in the United States under Rule 1 44A standards."[3]

In accordance with industry standards for underwriters of municipal bonds, investment bankers appropriately consider market conditions for waste-fuel projects.

Typically, an investor viewed by an issuer or underwriter as having sufficient resources, market knowledge and experience to understand and bear the risks involved in a particular investment.

"Qualified Institutional Buyer" (QIB): An entity to whom a security otherwise required to be registered under the Securities Act of 1933 may be sold without such registration under SEC Rule 1 44A. In general, a QIB must own and invest on a discretionary basis at least $100 million in securities and must be an insurance company, investment company, employee benefit plan, trust fund, business development company, 501 (c)(3) organization, corporation (other than a bank with net worth less than $25 million), partnership, business trust or investment adviser.[4]

---

[3] Lamb and Rappaport, *Municipal Bonds Book*, 276–294.

[4] See "Sophisticated Investor," *Glossary of Municipal Securities Terms*, Municipal Securities Rulemaking Board, supra note 2. The Municipal Securities Rulemaking Board (MSRB) was originally established by the U.S. Congress in 1975 to write investor protection rules and regulations of broker-dealers, banks, and other professionals. The MSRB publishes a glossary of official definitions of significant terms used in its public documents. "Sophisticated Investor" is an important MSRB definition because only such financial institutions' professionals or very wealthy investors are legally permitted to be sold these "nonpublic municipal securities" since "nonpublic securities" have potentially high levels of risk and financial complexity.

The bond offering materials disclose market conditions and competition in the market and the risks these factors posed. The bond offering documents will disclose whether a project is a functioning plant and, if not, when it will become operational. The bond offering materials will also provide extensive information on the project's fuel story.

Many municipal solid waste and power projects are developed from the start as publicly financed facilities with backstop guarantees from a state, city, county, or other public revenue authority. In other cases, private corporate developers often put their own capital up for several years to develop a new plant and guide it through the regulatory and political approval processes and technological development stages until it is ready to be underwritten by an investment bank and sold only to qualified sophisticated institutional investors.

Private corporate municipal financing projects very often encounter difficulties in qualifying for tax exemption under federal, state or local tax regulations, thus making operation more expensive in the earliest years of the project. States have "cap limits" for so-called "private purpose bonds" that, in addition to meeting a widely acknowledged public need, also have financial benefits for private development companies. Projects that have a mixed public and private purpose, such as privately developed and operated sewers and waste disposal facilities, power plants, hospitals, and low-income housing, are subject to such cap limits.

In an unenhanced municipal financing, such as this type of project, a leveraged debt payment structure involving interest-only payments in early years and a delayed bond principal amortization schedule is not unusual. In fact, it is relatively common. For example, the Colver waste coal power plant was an unenhanced municipal financing with this type of debt payment structure. The Kennedy International Airport cogeneration facility and the Stony Brook University cogeneration facility provide additional examples of municipal financing involving a "backloaded debt payment structure." Each has an interest-only debt payment schedule in early years and amortizes principal over more than 20 years.[5]

Private corporate municipal financing projects very often encounter difficulties in qualifying for tax exemption under federal, state, or local tax regulations, thus making operation more expensive in the earliest

---

[5] Stony Brook: SUFFOLK CNTY N Y INDL DEV AGY INDL DEV REV NISSE-QUOGUE COGEN PARTNERS FAC (NY)*, December 1, 1998, http://emma .msrb.org/IssueView/IssueDetails.aspx?id=MS195470 (accessed November 12, 2011); Kennedy Airport: KIAC Partners Cogeneration Project, December 9, 2010, http://emma.msrb.org/IssuerView/IssuerDetails.aspx?cusip=73358E (accessed November 12, 2011).

years of the project. States have "cap limits" for so-called "private purpose bonds" that, in addition to meeting a widely acknowledged public need, also have financial benefits for private development companies. Projects that have a mixed public and private purpose, such as privately developed and operated sewers and waste disposal facilities, power plants, hospitals and low-income housing, are subject to such cap limits. A private corporate municipal financing project may be structured as a leveraged lease, somewhat like car loans that have interest-only payments in the first year or two before loan principal starts to be amortized. Financial backers of waste-to-energy or power plants bonds have permitted private plant developers to customize debt payment structures with interest-only debt service in the early years and amortization of the principal payments starting after three or four years. This type of debt payment structure provides the private owner or developer with more cash with which to pay high-leveraged lease charges in the early years of the project, when the financial viability of new projects is most at risk. For example, the Foster Wheeler Passaic, Inc. waste-to-energy facility in Camden, New Jersey, was financed by bonds with this type of backloaded debt payment structure.[6]

In accordance with standards applicable to municipal underwriters, the bond offering materials disclose the debt payment structure. Institutional investors who purchased the bonds were expected to have their professionals (financial advisers, accountants, lawyers, etc.) evaluate the basic features of the bonds. Any professional reviewing the bonds should have noted that the first years of debt service were payments primarily of interest and that the bond offering materials contemplated the use of the debt service reserve to fund principal repayments at the end of the debt payment schedule. The primarily interest debt payments in the first years might also alert buyers that the project required a cash cushion in the early years. Prospective purchasers who were concerned about the fully disclosed debt payment structure of the bonds could have demanded modifications to that structure or declined to purchase the bonds.

Demands for specific changes in bond terms, yields, coupons, maturities, and debt payment structures are sometimes made by large financial institutions like plaintiffs. Such institutional investors are by far the largest buyers of tax-exempt bonds issued to finance projects in conjunction with private corporate developers and accordingly are in a strong position if they

---

[6] Passaic County, NJ, Pollution Control Financing Authority Solid Waste Resource Recover Rev. Foster Wheeler, Passaic Inc. PJ (NJ), December 20, 1990, http://emma. msrb.org/IssuerView/IssuerDetails.aspx?cusip=702760 (accessed December 9, 2011).

wish to negotiate bond terms or walk away.[7] Over 60 percent of the $80 million bond issue for one waste-to-energy project was purchased by one of the largest institutional investors in the municipal securities industry.

CALCULATION OF DEBT SERVICE COVERAGE

Projected operating results and the resulting estimates of debt service coverage were calculated using a calendar year, beginning January 1 and ending December 31, to show the project's projected income during that calendar year and its projected cash needs for all operating expenses and debt service.

The presentation of debt service on a calendar-year basis is one standard practice in the industry.[8] It is useful to present the information in this manner because it shows how much money the project must accrue for debt service that is due during a particular year in the course of its operations for that year. Debt service can also be presented on the so-called bond-year basis. This is also a standard practice in the industry.[9] Differences are explained by timing differences between the bond-year and calendar-year forms of presentation.

In the municipal securities industry, the term *independent engineer* or *independent feasibility study engineer* refers to a separate entity, not owned or controlled by the project's developer or underwriter, that is qualified to deliver an opinion as to the engineering feasibility of a project facility.[10] The firms that are qualified to serve in this capacity typically employ professionals with substantial engineering training and experience in evaluating solid waste facilities and power plants. A qualified sophisticated bond purchaser would be aware of and understand this concept of independence, which is commonly used in the municipal securities industry.

The underwriter must act in conformity with prevailing industry standards. This includes marketing the bonds only to qualified sophisticated institutional investors. The underwriter must have a reasonable basis to believe that the bond offering materials and other information available to

---

[7] Excerpt from Lamb expert witness report for a major municipal bond underwriter in litigation of nonpublic municipal issuance of private bonds to institutional purchasers for a major city's construction of a new replacement waste-to-energy power plant. See, for example, Deane Tr. at 277–279; Thornton Tr. at 67–72.

[8] Lamb and Rappaport, *Municipal Bonds Book*, 210, 236.

[9] See "Debt Service," *Glossary of Municipal Securities Terms*, Municipal Securities Rulemaking Board, supra note 2. Annual debt service refers to the total principal and interest paid in a calendar year, fiscal year, or bond fiscal year.

[10] Lamb and Rappaport, *Municipal Bonds Book*, 86–87.

prospective purchasers are sufficient to enable an investor to undertake a reasonable investment analysis of the bonds.

- The bond offering materials must disclose the market conditions for municipal solid waste, including the risks associated with potential fluctuations in fuel costs.
- The debt payment structure must be properly constructed, and the calculations of debt service coverage included in the bond offering materials were not erroneous.
- The independent engineer's independence is not subject to challenge under the standards of the municipal securities industry.

The duties and responsibilities of underwriters exist in accordance with customs, practices, and standards of care prevailing in the municipal securities business for the protection of investors. Underwriters generally perform "due diligence," essentially a review of relevant facts in order to have a reasonable basis to believe that key representations contained in municipal securities offering documents are reasonably accurate and complete and contain no misrepresentations or omissions of material facts that would make the offering documents misleading.[11]

The nature and scope of the due diligence performed by an underwriter are dependent on particular characteristics of the offering. Due diligence is an affirmative defense to a claim under Section 11 of the Securities Act of 1933 for registered offerings, and underwriters generally follow the same practices and protocols for unregistered offerings such as limited privately placed offerings of municipal bonds.[12]

For instance, see the following SEC release text:

---

[11] Lamb and Rappaport, *Municipal Bonds Book*, 226–227, 242–243. See also U.S. Securities and Exchange Commission (SEC), Municipal Securities Disclosure, Release No. 34-26 100, 53 Fed. Reg. 37778 (September 22, 1988) (SEC Release), Part III: "Municipal Underwriter Responsibilities."

[12] See, generally, SEC, Municipal Securities Disclosure, Release No. 34-26 100, 53 Fed. Reg. 37778 (September 28, 1988) (SEC Release), Part III: "Municipal Underwriter Responsibilities"; David S. Ruder, Chairman, SEC, Address before the Investment Association of New York, Underwriter Responsibilities in Municipal Bond Offerings after WPPSS (Sept. 22, 1988), www.sec.gov/news/speech/1988/092288ruder.pdf; and *Glossary of Municipal Securities Terms*, Municipal Securities Rulemaking Board, available at www.msrb.org/msrb1/glossary/glossary_db.asp?sel=d.

Municipal Underwriter Responsibilities

In the Commission's view, the reasonableness of a belief in the accuracy and completeness of the key representations in the final scope of the work performed include such matters as whether the bonds are general obligation bonds or revenue bonds; the nature of revenue sources used for repayment; and the purpose of the financing. For example, if the facility being financed is a major city airport, the due diligence would be vastly different than if the project being financed is new buses for a small town. In terms of absolute numbers, most municipal underwritings are for small issuers in relatively small amounts, but in dollar volume most of the $2.4 trillion municipal market involves large bond issues (tens or hundreds of millions of dollars) for states, cities and major revenue authorities.

The tasks that municipal underwriters perform also vary depending upon whether the bonds are being marketed widely to the public, or, conversely as here, are being sold though private placement to qualified sophisticated institutional investors. In the latter case, qualified sophisticated institutional bond investors are expected to perform their own professional due diligence before buying bonds. The underwriter brings together a working team of appropriate professionals to research, evaluate, and contribute information necessary to understand the project and its financing and to have a reasonable basis to believe that the bond offering documents make appropriate disclosures to prospective sophisticated investors.

The underwriter considers the professional expertise of the participants, meets frequently with the working group to review, evaluate and describe the material facts concerning the project. The underwriter asks the project developers to explain their assumptions, estimates, projections and specifications for the facility. Additionally, the underwriter reviews and questions the engineering firm charged with producing an engineering feasibility study of the project that will be bound together with the bond offering documents to be sent to potential purchasers. The underwriter also ascertains, with the assistance of counsel, that legal and regulatory requirements have been met, that the project is in compliance with applicable laws, and that the necessary contracts and other documentation are legally sufficient.

In a case like this, the various steps undertaken in advance of marketing the bonds to qualified sophisticated potential investors all share a common purpose—to permit the underwriter to have a reasonable basis to believe that such investors have available

appropriate information to undertake their own reasonable invest-
ment decision as to whether to purchase the bonds. Official state-
ment, and the extent of a review of the issuer's situation necessary
to arrive at this belief will depend upon all the circumstances.
Because of the varying types of municipal debt and extent of disclo-
sure practices, the Commission is not attempting to delineate spe-
cific investigative requirements in this release.[13]

The investment bank's commitments committee has to review and
finally approve the firm's underwriting and sale of the bonds. This was a
rigorous review process by the top managers of the investment bank.
Municipal bond sales managers and others at the investment bank participate in the private placement sale of the bonds to qualified sophisticated
institutional investors. The investment bank performs its role as underwriter in conformity with prevailing industry standards for an underwriter
of a privately developed waste-to-energy plant to be financed by municipal
bonds sold only to qualified sophisticated institutional investors.

It is expressly stated in the "Notice to Investors" in a Limited Offering
Memorandum:

No representation or warranty, express or implied, is made by the
Underwriters as to the accuracy or completeness of the information
contained in this [LOM], and nothing contained in this [LOM] is,
or shall be relied upon as, a promise or representation by the
Underwriters as to the past or the future.

INVESTMENT OPPORTUNITIES AT TROUBLED MUNICIPAL POWER PLANTS

For the next one to two years, today's low rates of interest due to the continuing U.S. recession should continue to provide various opportunities for
investors across the entire United States to take advantage of distressed
municipal power plants at potentially bargain rates. The U.S. Federal
Reserve Board's desire to keep interest rates low to enable the U.S. jobs
market and housing markets to recover from their deep decline, has led to
many financial funds and investors to be able to salvage troubled or

---

[13] SEC, Municipal Securities Disclosure, Release No. 34-26 100, 53 Fed. Reg. 37778
(September 28, 1988) (SEC Release), Part III: "Municipal Underwriter
Responsibilities."

previously non-financeable municipal waste-to-energy power plants at profitable rates.

As we stressed at the outset of this chapter, new favorable environmental legislation plus also new energy legislation at the federal, state, and municipal levels has made it possible to finance or refinance municipal waste-to-energy projects in ways that were not possible before. This is due to the fact that "waste-fuel power plants" can qualify for either municipal bond debt or the ITC/accelerated tax depreciation. Municipal bond interest rates today are too low, and therefore the U.S. ITC/accelerated tax depreciation is selected by most investors or funds today because it is currently a significantly more valuable financial benefit.

Any investors, partners, or investment funds that are interested in looking up any troubled waste-to-energy plant for potential bargains can start by going to http://IssueVie. This is the most useful and comprehensive municipal bond market access information database in the entire United States.

SUMMARY

Municipal bond-backed power plants can be an area for investment opportunity. Some of the very best investment opportunities in the entire energy industry today are in purchasing at fire-sale prices many abandoned or distressed municipal bond waste-to-energy power plants in our very low interest rate market. Why?

During the Great Depression of the 1930s, over 3,000 municipalities in the United States defaulted on their debt. However, in stark contrast to the many thousands of U.S. corporate bonds that became totally worthless, the vast majority of the defaulted municipal bonds across the United States came back to their full par value. The reason was simple: municipal bonds paid for essential public services, and as the economy recovered, there was more than enough to pay back the past bond debt and, in fact, to issue billions more new municipal bonds for the expanding U.S. economy.

The completion of today's abandoned or distressed waste-to-energy power plants will be needed soon because they fulfill essential public services.

In Chapter 6, we discuss energy storage.

Energy Storage

A penny saved is a penny earned.

—Benjamin Franklin

Nuclear power plants, coal power plants, gas power plants, and hydroelectric power plants all can produce energy seven days a week, 24 hours each day, and 365 days each year. Such daily, weekly, monthly, and yearly consistency of energy production from a plant is the required contract financial payment or guaranteed income stream demanded by most investors or bank lenders.

In contrast, solar power and wind power farms produce power only when the sun shines or when the wind blows. They usually stand idle most of the rest of the time. Often, they must spill off extra power if it cannot immediately be used in production. Science does not yet have giant new technologically innovative fail-safe batteries necessary to store huge amounts of wind or solar power overnight. Thus, by themselves, solar and wind power usually cannot be used for a financial loan guarantee covered by "liquidated damages" for a seven-day-a-week, 24-hour-a-day continuous stream of energy necessary for the vast majority of investors or bank lenders. Only with another extra backup source of power that can kick in each and every day, exactly when needed, can the power company ensure continuous power delivery to industrial, commercial, municipal, federal, or retail customers.

Also, it is vital to note that every such "auxiliary backup energy supply plant" fueled by gas, oil, or coal, which is needed to suddenly provide electric power in the pitch dark for a solar plant, or in the windless air for a wind energy farm, usually is much more expensive to run per hour than any normal standard power plants that are *always on*, running steadily. In short, there are high extra expenses to start and ramp up a second power

plant, as well as extra costs to shut down that auxiliary power plant at various times of the day or night due to light, dark, storms, or any and all
required unpredictable times, or due to mechanical or electric failure and
for unpredictable lengths of each day or night, or in hurricanes, hail, tornadoes, earthquakes, floods, fires, heat waves, or high seas—conditions that
may last days or weeks.

CHEAP ENERGY STORAGE—THE MOST VITAL GAME CHANGER IN THE WORLD

Solar power and wind power will not decrease the world's dependence on
traditional oil, coal, gas, or other fossil fuels until energy storage techniques
become widely available and cost less. "The unique challenge of the energy
sector is that electricity is the only product that is consumed within the same
millisecond that it is generated and delivered," said Terry Boston (CEO of
PJM, the largest electric transmission grid in the United States) at PJM's annual conference. "Only by finding a way to store energy can the potential of
renewable energies be fully realized."[1]

For many decades, the primary means of storing energy was called
pumped storage. The method of pumped storage was very basic and dates
back before the Romans. In Europe and the United States, old-fashioned
plumbing was fed from a holding tank of water placed on the rooftops of
many buildings throughout a city. The water was pumped up to the tank
on the roof at night when electric power cost least. Later, that water was
released and descended by gravity through pipes to bathrooms or kitchens
in each apartment. Old-fashioned toilets were flushed by pulling a chain on
a high water tank, which released the water by gravity.

Pumped storage for cities or factories today is simply a giant version of
the same gravity-fed system whereby millions of gallons of water in a lake
or reservoir is pumped uphill to a much higher lake, tank, or reservoir during the night, when the electric charges for pumping are the very lowest.
That stored water is later released to flow downhill to drive turbines, which
then drive electric generators.

---

[1] Terry Boston, speech on New Energy Storage Technologies,[2010 EPRI-PJM Storage Summit on April 20, 2010 "Energy Storage: The New Dimension" (accessed on
9/ 9/ 2010) A PDF of this document is on PJM's website under "Conferences." Electric Power Research Institute and PJM Interconnect co-conferences on Energy Storage 2010. PDF excerpts from conference speeches by U.S. Department of Energy,
FERC, Energy Industry professionals on energy storage.

One of the newest solar energy storage systems operates on an equally simple technology: the hot thermos bottle. For example, at solar power plants in Nevada, California, Arizona, and Europe, two solar power corporations—Abengoa Solar, a Spanish utility, and Solar Reserve LLC, of California—have both innovated new energy storage devices using pairs of huge tanks of boiling molten salt, which continues to deliver 6 to 12 hours of extra electricity, respectively, for heating, lighting, cooking, computing or for air conditioning, after the sun has set and the primary solar power plants have gone dark.

These pairs of tanks of molten salt are often 122 feet in diameter and at least 34 feet deep, which can hold and store 40 percent of the heat created by the power plant during the day.

These giant-sized solar plants are very expensive and use many high-intensity mirrors to focus the sun's rays on a liquid contained in tubes that are heated to very high temperatures. The liquid is used to boil water and create steam. By using a steam-turbine generator, electricity is produced. Abengoa's Solana plant in Arizona received a $1.45 billion loan guarantee for the 250-megawatt (mW) plant, which will heat 70,000 homes, and it has signed a 30-year contract agreement to sell electricity to Arizona Public Service Utility Company.[2]

Electricity from solar plants is especially expensive today at a time when natural gas prices have plunged, making natural gas–generated electricity cheap by comparison. Utilities that are under state mandates to buy more clean power, say solar or wind power, usually cannot justify the extra economic expense unless a special Department of Energy (DOE) government loan is provided to the plant, as in Abengoa Solar and Solar Reserve. In the meantime, Solar Reserve also has power sales agreements with NV Energy, Inc. and PG&E Corp., and expects to have the two plants in service by 2014. Each will cost $650 million to $750 million. Because of this golden promise of molten salt energy storage, this "concentrated solar energy" sub-industry has received large DOE financial grants, even in 2011 during the economic recession. Solar Reserve received a DOE conditional loan guarantee $737 million for a 110-mW plant in Nevada, and Solar Trust of America received a $2.1 billion DOE loan guarantee.[3]

The U.S. National Renewable Laboratory, part of the DOE, says, molten salt storage is "proven technology." Molten salt tanks retain 94 to 96 percent of the heat for later use when the sky is dark. These molten salt tanks' extended energy storage may make it possible for public electric

---

[2] Ucilia Wang, "DOE Offers $737M Loan Guarantee for SolarReserve Project," *GigaOM*, earth2tech (blog), May 19, 2011.
[3] Wang, "DOE Offers $737M Loan Guarantee."

utilities and also electrical grid operators to guarantee amounts of power on specific electric lines on specific time schedules. Governments, banks and investors backed those financings.

OPENING THE MARKET FOR HISTORIC ENERGY STORAGE FINANCING

November 10, 2011, was potentially a historic landmark day for U.S. energy storage. For the first time in history, both Republican and Democratic senators proposed a new Law called "The Storage Technology for Renewable and Green Energy Act of 2011" (STORAGE). This proposed law allows for a 20 percent investment tax credit up to $40 million to finance all energy storage technologies. If passed by Congress this is expected to jump start the entire energy storage industry and has a truly enormous potential to increase the reliability, security, and efficiency of this nation's electric grid.

Energy storage technologies help all forms of energy, whether it be natural gas, coal, nuclear, solar, wind, or some other green energy technology, in order to run more smoothly. Innovations in energy storage have been tested on the electricity transmission grid and it has proven this tax credit can help all kinds of developers to secure equity and debt financing for their projects.

Energy storage is truly a complementary technology which could expand and significantly improve the performance of all forms of electric energy. Prior to this proposed law, U.S. Federal Investment Tax Credits could only be used to finance energy that was to be immediately used on electric transmission lines of a particular grid.

The chairman of the U.S. Federal Energy Regulatory Commission, John Wellinghoff, had stated over a year ago that "The United States must end this old law in order to reduce the very significant risks to wind energy and solar energy, which absolutely require storage capacity to cover the hours of the day when "The sun does not shine and the wind does not blow." Less than two weeks before the Senate Storage Law was proposed, The U.S. Federal Regulatory Energy Commission (FERC) released its long anticipated Final Order No. 755 on frequency regulation compensation. The FERC determined that the current frequency regulation compensation practices of regional transmission organizations (RTOs) and independent system operators (ISOs), which do not account for the inherently greater amount of frequency regulation service provided by faster-ramping energy resources, are unreasonable and unduly discriminatory. The FERC's proposed order also opens up the possibility for a profitable new market in energy storage. Managers of wholesale electricity systems, such as RTOs and ISOs, must buy some power each day in order to maintain the frequency of the electricity over their systems.[4]

Regulating frequency by adding electricity to transmission systems at specific times and places can be complex. Usually, the faster and more accurately that current can be added, the more effectively it can rectify frequency problems on the grid. Frequency problems undermine the stability of the grid and in extreme situations can lead to electricity transmission system failures and blackouts. By law, managers of the systems must purchase electricity fairly and without discrimination among sellers. However, prior to the FERC's new order No. 755, most RTOs and ISOs thought this meant they had to pay the same amount of money per kilowatt for electricity used for frequency regulation purposes regardless of how quickly or accurately the seller of that electricity was able to supply it to the grid. This practice shut energy storage technologies, such as advanced batteries, out of the frequency regulation market. Although batteries can add electricity to the grid much quicker and more accurately than natural gas producers, the per kilowatt cost of battery power electricity is higher than by natural gas peaker generators which took longer to be ramped up and placed on the grid.

Now electric transmission system managers must consider the quality and speed of frequency regulation service in setting the prices. The FERC's new regulation puts electricity vendors who use energy storage systems into the frequency regulation business, and helps ensure them a profit. The need for frequency regulation service will rise as variable, renewable energy is added to the grid. Each 100 MW of wind energy requires about 3 to 5 MW of additional frequency regulation. Normal market forces will also encourage the addition of level-cost renewable energy to the grid and expand the size of the frequency regulation market. The big volume energy story of the next decade is widely expected to be the increasing use of natural gas for base load electricity generation. Ironically, this could be good news for wind and solar energy because natural gas prices historically have been highly volatile. While shale natural gas prices today are very low, natural gas prices will probably be volatile in the future. Utilities will hedge against that volatility by entering into long term supply contracts with renewable energy generators, whose fuel cost is usually fixed. The frequency regulation market is forecast to become large and highly profitable.

It is only by further developing these innovative energy storage technologies, like molten salt tanks, that new wind power, solar power, and other

---

[4] http://theenergycollective.com/jim-greenberger/68565/good-news-energy-storage -story.

green energies can be integrated into the existing electricity transmission systems. Because the electricity grids of both the United States and Canada, as well as those of Europe, keep growing and becoming more integrated, it is essential that these innovative energy storage technologies be developed, funded, and installed.

John Wellinghoff was a keynote speaker at the 2010 EPRI-PJM Energy Storage Summit where he forecasted the "broad adoption of storage technologies across the board, from flywheels and batteries for frequency regulation to compressed air and pumped hydro for wholesale and retail markets. The innovation of inexpensive energy storage technologies must provide reliable availability of energy at all times, but must also be better balanced in terms of costs throughout the year."[5]

CATEGORIES OF ENERGY STORAGE TECHNOLOGIES

Each of the different major types of energy storage technologies have very different lengths of time that each one is truly able to store amounts of electricity.

James McIntosh, director of renewable resources integration and grid architecture at the California Independent [Electric Transmission] Systems Operators (ISO), has spelled out in the following paragraph, how many minutes, versus how many hours, versus how many days, each of the major types of energy storage technologies can actually store electricity:

1. *Super capacitors normally are only able to effectively store electricity for up to 15 minutes.*
2. *Flywheels and Small Batteries can be effective for electricity storage for durations of 15 minutes up to one hour.*
3. *Stronger Batteries and Compressed Air Energy Storage can be effective for electricity storage for durations of one hour up to four hours.*
4. *Large Batteries, Pumped-Hydro Storage and Compressed Air Storage can be effective for storing electricity for four hours up to 24 hours.*

---

[5] "Energy Storage: The New Dimension," EPRI-PJM Energy Storage Summit Proceedings, April 20, 2010, www.pjm.com/committees-and-groups/stakeholder -meetings/symposiums-forums/~/media/committees-groups/stakeholder-meetings/ epri-pjm/postings/2010-epri-pjm-storage-summit-proceeding.ashx.

5. *Finally, only Compressed Air Energy Storage held in giant caves or abandoned mines, or Pumped Hydro Energy from giant reservoirs can be effective for storing electricity for several days.*[6]

Pumped Hydro Electricity Storage

In various nations, every night, when the cost of electricity is low, water from many hundreds of low elevation reservoirs has been pumped uphill to much higher elevation reservoirs. In the daytime when the price of electricity is high, water from all of the hundreds of much higher reservoirs is turned on in order to flow back downhill in many pipes through many turbines specifically in order to power the many electricity generators at the lower elevation power stations.

Underground pumped hydro storage facilities have been proposed as a new method of extending pumped hydro storage to a wider variety of geologic locations. A Canadian company, RiverBank Power, develops, constructs, and operates run-of-river and pumped storage hydropower facilities in North and South America. With offices located in Portland, Oregon; Logan, Utah; Rigby, Idaho; Toronto, Canada; and Lima, Peru; Riverbank is well positioned to become a market leader of hydropower in North and South America. Riverbank's development capacity of 1,100 MW of run-of-river and over 10,000 MW of pumped storage hydropower projects represents the largest hydropower development pipeline in the world [http://riverbankpower.com].

"Pumped hydro energy" is by far, the world's most widespread form of energy storage. However, it is expensive to acquire the real estate sites, and the rights, and obtain all the legal permits to dig two very large reservoirs at two different altitudes and then construct, transport, install, connect and test all of the pumping equipment between the two reservoirs to operate every night and day. Also, it can take up to a decade in lead time to decide, plan, finance and finally to develop a pump storage facility from scratch and to ensure it is fully operational.

Nevertheless, across America, this "pumped hydro electricity" accounts for 2.5 percent of total energy generation in the United States. As of 2011, this was the world's largest total amount of electricity energy storage capacity.

---

[6] "Energy Storage: The New Dimension," EPRI-PJM Energy Storage Summit Proceedings.

Compressed Air Storage

"Compressed Air Storage" is the second potentially huge volume electricity energy storage system in the United States and across the entire world. Once again, at night when electricity energy is cheapest, compressed air is pumped into giant caves or abandoned mines, then sealed shut and stored. At a later time, when peak energy needs are greatest and its cost is highest, that stored compressed air energy can be turned on and used to power a factory or used in the electricity transmission grid.

Because many giant caves or abandoned mines in older industrial regions are empty and left vacant, they are much cheaper to convert to compressed air storage than any other newly built container space. Also, on average, compressed air storage facilities only take three years to build compared to six to ten years it takes for pumped-hydro storage facilities to be built, and compressed air is rated at significantly less per kilowatt hour.

There are caves and abandoned mines in many regions across America in states along the Pacific coast from Washington to Southern California and from states along the Canadian border down to New Mexico, Texas to North Carolina.

Recently there has been a good deal of interest from states with large underground chambers left from their earlier industrial centers, such as Pennsylvania, Ohio, and the Great Lakes, as well as in areas in the Rocky Mountains where high wind velocity makes it possible to store wind energy.

A summary of Compressed Air Energy Storage CAES by Simon Pockley was posted on the Internet.[7] He spells out the basic science, history of key power plants and provides diagrams of underground and above ground CAES systems, plus photographs of the two major plants in the world.

For over 30 years compressed air energy storage has been used in plants in Huntorf, Germany and in MacIntosh Alabama in giant caves which have been successful. In fact, the Alabama CAES improved upon the German Huntorf design by incorporating an air to air heat exchanger to preheat air from the cavern with waste heat from the turbines. The plant has functioned with over 95 percent reliability demonstrating the viability of CAES technology in supplying ancillary services, load following and intermediate power generation.

The MacIntosh 110 MW CAES system was declared commercial on May 31, 1991. The cave is 220 feet in diameter and 1000 feet deep for a total volume of 10 million cubic feet. At full charge, the cavern is pressurized to 1100 psi and it is discharged down to 650 psi. During discharge,

---

[7] Simon Pockley, "Compressed Air Energy Storage (CAES)," May 19, 2008, www .duckdigital.net/Research/CAES_Assignment.doc

340 pounds of air flows out of the cave each second and it can discharge for 26 hours. Compressed air feeds a 100 MW gas-fired combustion turbine. In contrast to conventional combustion turbines, this CAES-fed system can start in 15 minutes instead of 30 minutes, uses only 30 to 40 percent of the natural gas, and operates efficiently down to low loads (about 25 percent of full load).

Today in America, the U.S. Department of Energy has financially backed a group of municipal utilities in Iowa and in nearby states in developing a new energy park to integrate a 75 to 150 megawatt wind farm with CAES technology. This Iowa project is expected to cost $200 million and operate by 2011 with the capacity to store 200 megawatts of power, enough for several days.

Innovative Storage Technologies

A large number of innovative companies are experimenting with a wide variety of new compounds to expand the range of new energy storage technologies. For example, a company called ESO plans to manufacture zinc air battery for large scale stationary storage applications. Aquion Energy is to manufacture low temperature sodium carbon batteries for large scale stationary appliances. Highview Power is to build Cryogenic Energy Storage Plants which use liquid air as the energy storage medium. GE plans to manufacture molten salt sodium nickel chloride batteries for mobile and stationary electricity storage. Sodium sulfer batteries manufactured by NGK Insulators are considered the only mature utility scale electrochemical storage device on the market. Yet, Natrion Corporation proposes to develop a new type of sodium sulfur (NaS) battery superior to NGK Insulators' molten salt version. Zinc bromine flow batteries for large scale stationary electricity storage are being developed. Zinc which is a component of several innovative energy storage firms' compounds is the world's fourth most common metal, so it is widely available if one of these energy storage firms succeeds globally.

U.S. REGIONAL MULTI-ENERGY STORAGE COLLABORATIONS

"New York State expects very significant growth in energy storage capacity for excess wind and hydropower, according to Robert Pike, director of market design for New York ISO. That is because New York already has 3 percent of peak wind load and about 7,000 mW of wind in the queue, which would bring it to 25 percent of peak load capacity. New York State has already installed a centralized wind forecasting system integrated as a 'dispatch tool that makes it possible to foresee wind changes and commit conventional wind storage resources around them.'"[8]

Pike noted that New York State has 1,500 mW of pumped hydro storage spread between Niagara Falls and Albany and "significant geological opportunities" (in caves) to add compressed air energy storage.

In other regions of the United States, there are various examples of quite similar coordination and collaboration between neighboring "grid systems." The top three wind-rich states" of Texas, Kansas, and Nebraska have a clear goal in developing their large scale bulk storage.

Jay Caspary, director of transmission development for Southwest Power Pool, sees "huge potential for off-system energy storage export. There are a lot of geologic formations in Southwest Kansas, Western Oklahoma and the North Texas 'Panhandle' that would support compressed air storage." Also, there has been a lot of interest in the potential for large pumped hydro facilities in the Missouri River Valley. "I see significant value to operations and markets having storage to help us deal with these intermittent resources."[9]

In a totally different type of regional multi-energy collaboration, "the University of Delaware, PJM, and other partners in the Mid-Atlantic Grid Interactive Car Consortium (MAGICC) have been running a demonstration project involving grid integration of plug-in electric vehicles (PHEVs). The vehicles feature a two–way flow of power. They're equipped to respond to a signal from PJM and discharge power from their batteries back to the grid for frequency regulation service."[10]

Electric cars are in use for only about an hour a day and idle the rest of the time. Dr. Willett Kempton who runs the grid integration of plug in electric vehicles demonstration project argues these PHEVs "present a "large storage resource." He described it as "storage at the end of the distribution system. We want to make use of all that storage that's out there." One of the advantages of grid-integrated vehicles is "using something that somebody else has bought and putting new controls on it. You're not paying for the battery: you're not buying the whole system."[11]

---

[8] "Energy Storage: The New Dimension," EPRI-PJM Energy Storage Summit Proceedings.
[9] Ibid.
[10] Ibid.
[11] Ibid.

FLYWHEEL TECHNOLOGY ENERGY STORAGE HAS THE LOWEST CYCLE-LIFE-COST

Flywheels have been used in factories and physics classes for over 170 years, as well as in various energy industries in order to generate ever faster energy production easily, inexpensively, and dependably. Flywheel technology is also being used for regulation service. Beacon Power was running and building 20-mW flywheel installations in New England and New York State. The plant in New England was in operation since late 2009. Construction was starting on the plant in New York State, and a second was in the approval process.

F. William Capp, president and CEO of Beacon Power, likened flywheels to "energy storage machines." They cost $25 to $30 million for a 20-mW flywheel installation—they have a long life, capable of 20,000 cycles over 20 years. Capp said the "cycle-life-cost is the lowest [of any] resource we're aware of."[12]

Older flywheel energy storage systems used a large steel flywheel rotating on mechanical bearings. Newer systems use carbon-fiber composite rotors that have a higher tensile strength than steel but are much lighter. Newer flywheel systems hover, spinning between magnetic bearings. They eliminate mechanical bearings maintenance and dangerous failures.[13]

Today's lithium ion polymer batteries can operate for only limited periods, such as 36 months. Yet, modern flywheels may potentially have the ability to spin indefinitely. Flywheels built as part of James Watt's steam engine have been continuously spinning for over 200 years. Working models of ancient flywheels have been discovered in Asia, Africa, and Europe, which were used for making pottery, milling of grains, and sharpening weapons. In short, of all energy storage mechanisms, flywheels reach high speeds more quickly, and they can run potentially indefinitely, so they are being experimented on widely throughout the world.

The vital weakness of flywheel technology for energy storage is the tensile strength of the physical material of the rotor. When the tensile strength of the flywheel is exceeded, the flywheel will shatter, causing a "flywheel explosion" in which there are fragments of metal that are lethal. As a result, modern flywheel rotors made of carbon-fiber composite polymers become clouds of dust, not steel shards if they experience a flywheel explosion.[14]

However, Beacon Power, a key manufacturer of flywheel based energy storage systems was forced to file for bankruptcy on October 31, 2011

---

[12] "Energy Storage: The New Dimension," EPRI-PJM Energy Storage Summit Proceedings.
[13] www.distributedenergy.com/may-june-2007/primer-flywheel-technology.aspx.
[14] M. M. El-Wakil *Power Plant Technologies* (New York: McGraw-Hill, 1984).

because its costs were high: much higher than the current extraordinarily low cost of shale gas.

Supercapacitors or Ultracapacitors

Supercapacitors are another key energy storage device which now have become vital to the automotive, aircraft, and new electronic and computer devices industries. They are being innovated upon to be used in energy storage devices for power plants. A supercapacitor charges in seconds, has no danger of overcharge, have high rates of charge and discharge but traditionally they could only provide very short term energy storage.

However, supercapacitors today have high cycle efficiency (over 95 percent or more). In fact, new supercapacitors producers claim they can cycle millions of times = 10- to 12-year life. Some of the newest supercapacitors are tiny and also biodegradable and can be powered by body fluids and used in medical devices to repower and discharge energy inside the human body.

Supercapacitors disadvantages are that their linear discharge voltage prevents the use of the full energy spectrum and their low energy density—typically they hold one fifth to one tenth the energy of an electrochemical battery. Supercapacitor cells have low voltages—serial connections are required to achieve higher voltages. Also, voltage balancing is required if more than three capacitors are connected in series.

Historically, supercapacitors were initially used by the U.S. military to start engines of tanks and submarines. Most applications of supercapacitors today are in small appliances, handheld electronics and hybrid electric vehicles.

Most hybrid vehicles use 42 V supercapacitors. General Motors developed a pickup truck with a V8 engine that used the supercapacitor to replace the battery. The efficiency of the engine rose by 14 percent. The supercapacitor supplies energy to the alternator. Toyota has developed a diesel engine using the same technology and it is claimed to use just 2.7 liters of fuel per 100 km.

From the standpoint of supercapacitors' capability for being used in power plants, they need batteries to store the energy. They are basically used as a buffer between the battery and various types of power devices. UPS battery backups provide power protection for all kinds of electrical equipment. As a result, supercapacitors are being merged with batteries into a kind of "hybrid battery." Super capacitors can be charged and discharged hundreds of thousands of times, which basic batteries cannot do. Because the price of supercapacitors is expected to decline, supercapacitor manufacturers argue that virtually anything now powered by batteries could be improved by a supercapacitor energy supply. They can be made in any size. Their light weight and low cost make them a valuable add on to batteries to develop new types of energy storage devices.

Today, supercapacitors are used for energy storage for short periods of time in power plants and other energy storage situations. In order to expand their use in power plants and sizable energy storage supercapacitors must be further developed to merge with batteries on a large scale.

SUMMARY

Since renewable power plants don't supply power 7 days a week/24 hours a day, their future success will be determined by energy storage: The new energy storage law's investment tax credits & FERC's new rule help. Many new cheap energy storage technologies could be the most vital game-changers in the world. Pumped hydro energy storage, which is the oldest and most widely used method in the world for centuries, continues to be the leader. But, compressed air storage in abandoned mines and caves is now being built in various parts of the United States and the world and it is a much cheaper and quicker built storage technology to provide energy storage for whole states and regions of a nation. Solar energy has now fully developed giant molten salt tanks next to the solar plants, so these are now a "U.S. government proven storage technology" to power the generators on solar plants for 8 to 12 hours when the sun is not shining. Yet, solar plants today can cost billions of dollars.

Likewise, new electric power centers containing many lithium ion batteries are being used to supply temporary electricity for power plants in the nation of Chile and are beginning to be used in the United States. Electric car fleet parking depots are also now being used to supply energy for power plants. Flywheels, which last for several decades or even potentially over 100 years of continuous operation, are being introduced into energy power storage systems as the longest-lasting form of energy storage, but they too, are expensive.

In short, various innovative energy storage technologies are now expanding the range of types of energy that can become connected to the power grids across the United States and other regions and nations. This enables more widespread geographic and collaborative energy development that could ultimately lower the regional, national, and international cost of energy, widen the availability of energy, and also improve the civilian and military security of energy grid systems.

However, the key question remains whether shale natural gas, currently the lowest cost energy, will prevent vitally necessary new investments in wind, solar panel, solar thermal, wave, tidal, geothermal and all the other alternative energies to develop and prevent funding all new energy storage technologies they require.

In Chapter 7, we discuss shale natural gas.

Shale Natural Gas and Its Effect on Renewable Power

Success is half luck and half brains.
— Kemmons Wilson, founder of Holiday Inn

As was described in the chapter on coal, renewable power projects operate in world of competing technologies and fuels. As was also mentioned, recent technological advances have allowed large amounts of U.S. shale gas to be cheaply developed. These new technologies include horizontal drilling and hydraulic fracturing. One company, Mitchell Energy, worked out how "slickwater" fracturing combined with horizontal drilling could free gas from dense shale rock previously uneconomical to develop. Last year, the firm's founder, George Mitchell, received the Gas Technology Institute's lifetime achievement award.[1]

FRACKING

Gas production rates are dependent on porosity or the gaps between rock particles, and permeability, or the ability of fluids to move through the rocks due to the connectedness of those gaps. Fracking or the forcing of fluids at high pressure into a wellbore, will create cracks in hydrocarbon-bearing rock that will allow the good stuff to flow out.[2] A recent concern is that the wastewater from shale drilling operations has a high radioactivity level.

---

[1] H. Jenkins, "Listening to the Shale Revolution," *Wall Street Journal*, February 5, 2011.
[2] J. Dizard, "The Shale Gas Fairytale Continues," *Financial Times*, July 18, 2010.

The concern is that this water is being sent to wastewater treatment facilities that are not able to properly treat it and is ultimately being released into rivers and streams. This could cause problems with drinking water supplies. Since the fracking process is new, regulatory agencies are not used to dealing with issues of this type.

U.S. shale gas helps with security of energy supply in that it is not imported from an unfriendly country. Foreign firms are investing in U.S. shale gas plays in order to learn about the business. However, this currently cheap natural gas makes it difficult for renewable power plants and demand-side management to compete in the electric power market. Regulators that had encouraged electric utilities to sign power purchase agreements with renewable projects are now concerned that the future price for power from renewable projects will be too high over the long term. This low price for natural gas is also making it difficult for coal power plants to compete.

There are also potential large shale gas reserves in China and coal bed methane and tight sands gas in India. Poland alone is estimated to hold shale gas reserves equal to half of Europe's existing conventional reserves—a fact already altering the strategic balance between Europe and its soon-to-be-former energy overlord, Russia.[3] On February 28, 2010, the Department of Energy's (DOE) Energy Information Administration (EIA) reported that the United States produced 21.57 trillion cubic feet (Tcf) of natural gas thanks to shale gas production. This is just short of the 1973 record of 21.73 Tcf.[4]

NEW ATTITUDES IN NATURAL GAS

This recent abundance of natural gas is a major change from the restriction on the use of natural gas to generate electricity under the Fuel Use Act and Project Independence during President Jimmy Carter's administration. A large number of nuclear and coal plants were partly spawned by the Fuel Use Act, which was repealed in 1987. In July 1989, the Natural Gas Wellhead Decontrol Act was signed into law. This act allowed for the complete decontrol of natural gas prices by January 1993. In the early to mid-2000s in the United States, there was a concern that a dependence on natural gas would change America from being dependent on one Organization of Petroleum Exporting Countries (OPEC) to another. This new OPEC would

---

[3] H. Jenkins, "Listening to the Shale Revolution."
[4] Matt Day, "Record U.S. Natural-Gas Output Likely to Continue in 2011," *Wall Street Journal*, March 2, 2011.

be countries that had large deposits of natural gas that would be shipped to the United States.

Renewable projects face a situation where they have to compete against natural gas–fired combined-cycle power plants since they are able to get permitted and built cheaply. This creates a classic tactics versus strategy challenge. Natural gas plants can be built quickly, but there is no long-term energy plan for the United States. Most politicians are in office for only four years. This is not a long enough time to set overall energy policy. Developers and regulated utilities also understand that natural gas–fired power plants will be required to back up renewable power plants. Unlike coal plants, natural gas turbine engines can be stopped and started relatively quickly and have a better ability to curtail their output.

In the recent past, the price of natural gas was considered very volatile. It was possible to fix a delivered price of natural gas using hedges. This involved cost and potential counterparty risk. Industrial customers would be concerned about using natural gas for their manufacturing processes due to this concern. Now industrials may find that there is a new shale gas deposit in their backyard or the local area.

Liquefied Natural Gas

A number of liquefied natural gas (LNG) terminals have been built in the United States to import natural gas from places like Angola, Qatar, and Trinidad. In these natural gas–rich countries, it is not uncommon for the production cost of the gas to be as low as $1/million British Thermal Units (MMBtu). As a result, the economics of shipping this gas to foreign markets work. Unlike the United States, these countries have not developed pipeline and overall natural gas infrastructure that allows for the widespread use of this domestic gas. These export countries have also understated the true cost of natural gas for domestic consumption and have only recently raised the price.

The production cost of U.S. shale gas is at least $4/MMBtu and higher in most shale deposits. The DOE EIA feels that a gas price of $4/MMBtu acts like a floor for natural gas prices. At a price less than $4/MMBtu, there is a switch to natural gas to back out coal. The DOE EIA feels that it is hard to argue that natural gas will stay below $4/MMBtu. The Marcellus Shale is said to be economic at a gas price of $4/MMBtu. The Haynesville Shale is economic at $5/MMBtu. Some have argued that this price covers only finding and development costs and nothing else. There is also a concern that more stringent future permitting requirements could also raise the cost of production. Most conventional natural gas needs $7 to $8/MMbtu to be economic. Shale gas producers have stated that the price of gas has to get to

TABLE 7.1 Five Largest Shale Plays

Barnett	TX
Fayetteville	AK
Woodford	OK, TX, NM
Haynesville	LA, TX
Marcellus	PA, NY

$6 or $7/MMBtu to support the last marginal MMBtu that the United States needs.

In the United States, says the *Oil & Gas Journal*,

> *Companies generally can develop shale plays located in the U.S. Midcontinent and East, where most land is owned privately, with minimal political wrangling. The fact that shale developments can cover entire counties means that royalties are spread among thousands of individual landowners, often aligning them with operators.*[5]

The five largest shale plays are shown in Table 7.1.

The overall economics of individual shale gas wells may be better than is widely thought at the current low gas prices since they may also produce either natural gas liquids or oil. In the current market, both of these products are more valuable than natural gas. Natural gas liquids are used as a feedstock for chemicals. This also helps explain why shale gas producers can continue to be profitable even at very low prices for natural gas. The value of natural gas liquids and oil could be reduced be a lack of infrastructure to move them to market.

COST OF PRODUCTION

Since this cost of production is higher than for markets like Qatar, Algeria, or Russia, the economics of exporting shale gas to other markets is difficult. Since U.S. shale gas is mostly "stranded natural gas," this will help keep the

---

[5] Scott Stevens and Vello Kuuskraa, "Special Report: Gas Shale—1: Seven Plays Dominate North America Activity," PennEnergy, www.pennenergy.com/index/petroleum/display/8128977500/articles/oil-gas-journal/volume-107/Issue_36/Drilling___Production/Special_Report__GAS_SHALE_1__Seven_plays_dominate_North_America_activity.html.

price low in the United States for the near term. U.S. law allows for mineral rights to be sold or leased by landowners to shale gas developers, which makes development easier. There is an argument that industries that require a large amount of natural gas might be encouraged to move to the United States.

LNG projects are financed on a project finance basis and require long-term offtake contracts. All of the LNG facilities currently located in the United States are based on importing and not exporting natural gas. It would be expensive and time consuming to reconfigure these terms to export natural gas. It will be difficult to obtain a long-term supply contract with shale gas suppliers who are concerned about capturing future price increases. Shale gas wells have a steep decline curve, and this would be a challenge for an LNG terminal. LNG terminals require a 20-year supply of natural gas to match their offtake contracts. The quick decline curve of a shale gas well would be hard for an LNG developer to contract with. Although some natural gas is a long-lived resource, peak production generally exists for a relatively short period and then declines throughout the remainder of its life. This situation is especially true for shale gas. In the Marcellus Shale, there is a 50 percent decline in natural gas production by year two. As a comparison, coal is a long-lived resource and is typically sold under long-term contracts.

Asian countries that import LNG typically agree to long-term contracts at 85 to 90 percent of the price of oil on a Btu basis. If one assumes an oil price of $90/barrel, this calculation becomes:

$$\$90/\text{Barrel} \times 1\,\text{Barrel}/5,848,000\,\text{Btu} \times 1000000\,\text{Btu/MMBtu}$$
$$= \$15.39/\text{MMBtu}$$

At a multiplier of 85 to 90 percent, this becomes $13.08 to $13.85/MMBtu. As a comparison, the U.S. Henry Hub natural gas price is currently in the range of $4/MMBtu or a factor of 3.27 to 3.46 times lower. The overall cost of construction has increased for new LNG terminals. An annual tonne of gas liquefaction has increased from $400 to $1000 over the past decade.[6]

The Marcellus, with 489 Tcf of recoverable reserves, is second in the world in size to the South Pars/North Dome, which is located in Iran and Qatar with 1,235 Tcf. This is very interesting since conventional thinking was that few if any elephant natural gas reserves remain to be discovered. According to the DOE EIA, the 2009 U.S. natural gas demand is 23 Tcf,

---

[6] J. Dizard, "Pitfalls of the U.S. Cheap Gas Habit," *Financial Times*, February 13, 2011.

and the Marcellus alone would meet the U.S. need for over 21 years. The U.S. gas resource potential is 1,836 Tcf,[7] which at 2009 consumption levels is enough gas for 80 years.

Compressed natural gas (CNG) vehicles make sense for urban fleet vehicles (e.g., delivery vehicles and buses). In addition to limitations with batteries, there is an argument that the electric grid will not be able to handle the load from a large implementation of electric cars at peak times. There is currently limited infrastructure for natural gas vehicles. Honda is the only consumer auto manufacturer to produce a car that runs on CNG.

CNG engine manufacturers require natural gas supply of a high purity. This can require natural gas utilities to stop using their propane air facilities. Propane air facilities help natural gas utilities to extend their winter natural gas supply. In some areas, refinery gas or ethanes are mixed in with the natural gas supply and will also have to be removed from the gas grid. This can be a technical and environmental challenge that can require study time and capital expenditure.

Natural gas–fired power plants also produce about 50 percent less carbon dioxide (CO_2) emissions than a coal-fired power plant. This has become less of an issue recently. In fact, the Chicago Climate Exchange (CCX) closed down on November 28, 2010. It had stated that it was "North America's only cap-and-trade system for greenhouse gases." At its peak in May 2008, CCX was trading 10 million tons of carbon permits per month. The price of carbon hit a high of $7.40/ton in mid-2008. The market collapsed in 2009 when prices fell to $1/ton. The lack of success of the CCX is related to the failure of Congress to pass a mandatory cap on carbon emissions.

There will be a number of interesting opportunities with natural gas storage. Even though the current price for natural gas is low, there can be volatility in natural gas prices on a daily and/or seasonal basis. A number of natural gas storage facilities are being developed to take advantage of this situation. Natural gas–fired power plants are required to backup renewable power plants and might have to start running on short notice. The large amount of natural gas that they require will also have to be stored. Since renewable power projects can go offline quickly and unexpectedly, it is unlikely that a sufficient amount of natural gas would be available in the local transportation grid to run a gas turbine engine for any length of time. There is a natural gas storage project under consideration in the state of

---

[7] "Potential Gas Committee Reports Unprecedented Increase in Magnitude of U.S. Natural Gas Resource Base," Colorado School of Mines, June 18, 2009, www .mines.edu/Potential-Gas-Committee-reports-unprecedented-increase-in-magnitude -of-U.S.-natural-gas-resource-base.

TABLE 7.2 Shale Gas Well versus Traditional Well

	Initial Production	Reserves	Capital Cost
Old well	100 Mcf/day	0.3 Bcf	$0.5 M
New well	4,000 Mcf/day	4–6 Bcf	$4-5 M

Colorado in order to back up the large number of wind projects that are currently under development. Power plants typically nominate the amount of natural gas that they use the day before they actually dispatch or run. Due to the uncertainty of the dispatch of a wind project, this level of information is unknown.

The concern is that if shale gas is more expensive than first considered or if the wells have a faster decline curve than first calculated, a country could be exposed to high energy cost. There is no longer-term operating history with shale gas. The science of modeling shale gas deposits is getting better, but is still underdeveloped given the capital and policy commitment to the industry.[8] In the United States, it is next to impossible to develop a new coal plant, and a number of smaller coal plants may be forced to shut down. This also increases the overall U.S. exposure to potential natural gas price and supply risk.

Table 7.2 helps to illustrate the large cost and production differences between a shale gas well and a traditional well.

It is harder for consumers of natural gas to respond to cheaper prices. It is easy for consumers to just drive further in order to consume more gasoline as a result of cheaper oil. Residential demand for natural gas has been stagnant for years with increasing efficiency of gas heating systems, while demand from energy-intensive industries such as chemicals and metals manufacturing was hit by the recession and declines in the U.S. manufacturing base.[9]

There has been a concern about the large use of water for shale wells. One study from Range Resources finds that all of the wells in the Marcellus in Pennsylvania use 60 million gallons of water per day. As a comparison, all of the power generators in Pennsylvania use 5,930 million gallons of water per day.

---

[8] J. Dizard, "Debate Over Shale Gas Decline Fires Up," *Financial Times*, October 10, 2010.
[9] Day, "Record U.S. Natural Gas Output Likely to Continue in 2011."

SUMMARY

Shale gas has challenged the economics of coal, renewable power, and demand-side management. The recent production of shale gas has dramatically changed both the fossil and renewable energy markets. Unlike other inventions in the energy market, shale gas is immediately relevant since it fits in the existing energy pipeline and power equipment infrastructure.

Chapter 8 discusses different types of solar power projects and their economics in a world of abundant shale gas.

Solar PV and Solar Thermal Power Plants

Logic will get you from A to B; imagination will take you everywhere.

—Albert Einstein

Like wind power projects, solar power projects face the challenge of not being a seven-day-a-week, 24-hour-a-day, much less a 5 × 16, resource. The economics of solar projects are also dependent on tax credits and/or tax equity investors in general.

THE ECONOMICS OF SOLAR POWER

Wind projects are paid a renewable energy credit (REC) for each megawatt hour (mWh) of power produced. Solar projects also generate REC credits called solar renewable energy credits (SRECs). The current low price for natural gas and the lack of a meaningful tax on carbon also makes it tough for solar power projects to compete in the United States. Solar photovoltaic (PV) projects currently represent less than 1 percent of the nation's power assets.[1] According to the Department of Energy's (DOE) National Renewable Energy Lab (NREL), about 0.1 percent of the earth covered with 10 percent efficient solar cells could provide the whole world's energy needs. The reality would be the siting, intermittency, coordination, and transmission and distribution of all of these disparate solar cells.

---

[1] Steven Andersen, "Solar Dawn," *Public Utilities Fortnightly*, March 15, 2011.

Solar projects also face the challenge that there is no national renewable portfolio standard (RPS) in the United States. As with wind, many states have already met their RPS requirements. States typically have a different category for wind and solar power RPS standards. The capital cost estimates for solar PV projects are in the range of $3,800 to $5,000/kW. Solar PV projects have a capacity factor only in the range of 21 to 31 percent. U.S. states have been unrealistic on the potential size of their RPS standard. Massachusetts claimed to be relying on future power plants to be located in the state of Maine in order to meet its unrealistic RPS standard.

Unlike wind projects, solar projects tend to produce power at times when it is required by the grid. Utilities have been willing to agree to higher-priced power from solar projects since the power tends to be available during on-peak times. A utility could compare the cost of power from a solar project to that of a peaking gas turbine engine. There have been discussions on having a grid based on solar power during the day and wind in the evening. This concept misses the point that solar and wind power projects are not dispatchable like a gas or coal fossil power plant and the weakness of today's battery technology.

In the current market, solar developers are having a difficult time obtaining long-term, financeable offtake contracts for the SRECs that they produce. Only local utilities can purchase SRECS; they are currently not traded by Wall Street firms and other energy brokers. In order to obtain nonrecourse project financing, lenders want power project to have long-term defined prices in their offtake contracts. A utility can get comfortable about taking the energy from a solar project over a long period since it has a regulated rate base. SRECs, however, are not part of their rate base, and, like other parties, they are concerned about a drop in SREC pricing. Utilities also face the same problem that independent operators face—the limited liquidity in the SREC market makes it difficult to hedge.

FINANCING TECHNIQUES

The internal rate of return for solar projects can also be below that of an investment in the first or second lien of a distressed natural gas power plant. Similarly, solar projects produce a large number of tax benefits that typically can't be used by small developers. Solar projects also have been using partnership flip structures and leveraged leases. Some developers have been taking advantage of solar power equipment built in China. China has been willing to subsidize solar PV technology. Competition with China in this case is not an option.

Solar projects have more experience with using leveraged leases since they always qualified for the investment tax credit (ITC). As mentioned, it is only recently that wind projects qualified for the ITC. Wind projects had previously only qualified for the production tax credit (PTC). The ITC can pass through a lease, whereas the PTC can't be used in a leasing structure. The PTC also requires investors to take performance risk and to forecast if they will have enough tax appetite for the next 10 years. Leveraged leases can provide 100 percent financing and avoid the capital account and outside basis issues in a partnership. At the end of the lease, the lessee is typically provided an option to buy the lessor's position at a market price. The expenses of putting a lease together is high and will typically work on larger projects.

Solar developers are also taking advantage of the 1603 cash grant, which was extended until the end of 2011. This strategy entails taking the cash grant for the 30 percent ITC and then carrying forward the five-year modified accelerated cost recovery system (MACRS) depreciation as the project generates earnings before interest, taxes, depreciation, and amortization (EBITDA) to use it. Like wind projects, the DOE loan guarantee program also expired for solar projects in September 2011. A number of solar projects are teaming up the cash grant with leverage, which could create some future restructuring opportunities. In the current environment in Washington, the cash grant was not extended past the end of 2011.

Solar projects have even less operating history than wind projects. The argument that solar developers are making to continue the grant program is that it will help with the continued commercialization of solar projects. It can be argued that the Public Utilities Regulatory Policy Act (PURPA) encouraged the development of numerous natural gas power plants, which helped further commercialize gas turbine combined-cycle power plants in cogeneration applications.

It has also been considered that a solar power plant could replace a natural gas turbine operating as a peaker. The constraint that the solar plant would have is size. The 21-mW solar PV plant located in Blythe, California, requires 200 acres! This project will generate only 45,000 mWh per year or will be available to generate electricity 24.5 percent or 2,146 hours per year. The peak demand time for electricity is during the summer months of June to September. Based on 52 weeks per year and 12 months per year, there are (52/12) or 4.33 weeks per month. Since on-peak power is calculated in the California market based on 6 days per week and 16 hours per day, this converts to 96 hours per week. Multiplying 96 hours per week by 4 summer months and 4.33 weeks per month equates to 1,663 hours. This is only 77 percent or 1,663/2,146 of the hours that are available to be produced from the Blythe project.

If there were a local power requirement for over 100 mW, this would be easy to meet with a new gas turbine engine. As a reference, a 240-mW solar PV power plant under construction in Arizona will require 2,400 acres of land. Based on the conversion of 1 square mile equals 600 acres, this converts to 4 square miles. A solar PV facility of this size would require and unmanageable amount of property. If the sun didn't shine or if there were a lot of clouds, the solar facility would have to pay liquidated damages for lack of performance and could wipe out any profits. There is also limited operating history for solar power projects, and long-term availability, operating cost, and reliability are untested. Solar thermal projects are tough to build and project finance due to the difficulty of obtaining a lump-sum, date-certain, fixed-priced contract with liquidated damages. As with integrated combined-cycle coal plants, it is difficult to obtain performance guarantees from the equipment suppliers of solar thermal power projects.

There have been many solar projects developed in Europe due to the feed-in tariff program. In fact, 85 percent of the installed global solar market is in Germany, Spain, Japan, and the United States.[2] These projects didn't have to deal with tax issues or renewable credits. They were paid a high price from the start of operations. As a result of the recent economic crisis, a number of the European countries have released that they could not continue to pay these very high prices for power. Spain set the price for the feed-in tariff too high and created a solar boom. Solar power in Spain grew from 400 mW in 2007 to 2.5 gW in 2008.[3] Spain has started the process of attempting to reduce the price for power for solar and other renewable power projects. Germany has reduced the price for power for new solar projects. Like wind, solar power projects produce a small amount of EBITDA and a reduction in tariff could cause a default situation for a project with debt and or high operating costs.

THE TECHNOLOGY

There are two main types of solar technology: solar photovoltaic and solar thermal. Solar PV includes both thin film and crystalline. Both thin film and crystalline has both utility scale and commercial rooftop versions.

Solar PV technology has the following attributes:

- Uses "global isolation"—can be installed anywhere.
- Clouds = immediate loss.

[2] http://files.eesi.org/Clavenna_071108.pdf.
[3] www.kinesis.org/pdfs/KINESIS_MONITOR_15_MAR2010.pdf.

- No economical large-scale storage option.
- Unattended operation.
- Approximately 10 to 15 percent solar to electric.
- Mostly mature technology.
- Can be fixed orientation: no moving parts.
- Modular—same technology for 30 kW, 3 mW, and 100 mW.
- Simple permitting and development.
- Low water usage.
- Rapid installation time.
- Costs are well known.[4]

Solar thermal technology has the following attributes:

- Uses "direct isolation"— viable only in desert Southwest.
- Thermal inertia.
- Commercial thermal storage for some systems.
- Typical power plant operations and maintenance (O&M).
- Approximately 10 to 15 percent solar to electric.
- Many precommercial technologies.
- Must track the sun in one or two axes.
- Typically requires large systems for economy of scale.
- More difficult permitting.
- Wet cooling (most efficient) requires substantial water.
- Longer installation.
- Costs can be hard to estimate.[5]

Solar projects that are located in sensitive areas can also face local opposition issues just like fossil-based power plants. Solar projects located on federal land require permits from the Bureau of Land Management (BLM). In addition to NIMBY (or not in my back yard), opposition to solar and other power projects also include the following:

- LULU: locally undesirable land use.
- NIMTO: not in my time of office.
- NIMEY: not in my election year.
- BANANA: build absolutely nothing anywhere near anybody.
- NUMBY: not under my backyard.

---

[4] Black & Veatch Corporation, Energy Seminar in New York City, November 3, 2010.
[5] Black & Veatch Corporation Energy Seminar.

Due to these issues; overall cost; and engineering, procurement, and construction (EPC) contracts, a number of solar projects recently have been canceled. In late January 2011, developer Solar Millennium canceled its 250-MW parabolic trough solar plant near Ridgecrest in California's Mojave Desert. In December, Southern California Edison canceled a 20-year 663.5-mW power purchase agreement (PPA) it had with a parabolic dish project. In June 2010, a 107-mW solar and biomass project in Fresno County, California, which had a PPA with Pacific Gas & Electric, withdrew its license application. Another 290-mW solar thermal project was canceled in late 2009 after the EPC stated "unexpectedly high supply base costs" as well as overall size and risk profile of the project.[6]

Solar radiation components are based on direct and diffuse insulation and reflected or albedo radiation. Global radiation is the sum of all of the following.[7] Solar PV plants used global radiation. Solar developers use the NREL National Solar Radiation Database to help in the development of their projects. Typically, only solar thermal and very large solar plants require onsite measurement. With PV technology, sunlight is converted to direct current (DC) electricity in a semiconductor material. An inverter converts the DC electricity to alternating current (AC). With solar thermal, sunlight heats a fluid. A turbine or engine is used to convert heat to electricity. Solar thermal uses mirrors to concentrate sunlight.

SUMMARY

Solar power has two main technologies and faces a number of challenges, including a large land requirement. The DOE loan guarantee program and the cash grant have helped to support the recent growth of solar power plants. Like other renewable power plants, solar projects face extremely tough competition from large amounts of inexpensive and plentiful shale gas.

In Chapter 9, we describe wind power—the other most developed renewable power technology.

---

[6] Steven Andersen, "Solar Dawn."
[7] Black and & Veatch Corporation Energy Seminar.

Wind Power Plants

Stay hungry. Stay foolish.
>   —Steve Jobs, Apple Computer founder and CEO

Wind power projects face tough competition from other fuels. This is especially true from the current low price of natural gas. The overall demand for power is depressed due to the slow economic recovery and the fact that numerous natural gas power plants were built in the past 10 years. Wind projects are also dependent on tax subsidies, which are currently difficult for most developers with no or limited tax appetite to use. The extension of the cash grant has provided a short-term fix to the lack of tax appetite among tax equity investors. Each state also has varying degrees of a support for wind via renewable energy credits.

PROJECTS OVERVIEW

Due to the large size of the United States, it is not possible to reliably power the entire grid with renewable power plants. Unlike the United States, Iceland, with a population of only 318,500 people, hit the renewable power lottery due to its geography. This concept was further developed by John Dizard:

> The hydroelectric and geothermal power stations in Iceland generate five times the national requirements in a non polluting way. Power is exported as end products of an energy intensive process aluminum and ferrochrome.[1]

---

[1] J. Dizard, "Iceland Is Hotter than You Might Think," *Financial Times*, October 15, 2010.

According to the Department of Energy's (DOE) Energy Information Administration (EIA), renewable energy provided for 8 percent of the U.S. energy consumption in 2009. Of this 8 percent, wind made up 9 percent. In 2009, 38,610 megawatts (mW) of wind power plants were installed globally, and in 2010 this fell to 35,802 mW. The average size of wind power projects has also increased from 50 mW in 2003 to 150 mW in 2009.[2]

Onshore wind projects have an all-in cost in the range of $2,200 per kilowatt (kW). This includes capital cost, development cost, and interest during construction. The dispatch of a wind project can range from only 30 to 42 percent. In comparison, a fossil power plant can operate at an availability of over 90 percent if required by the electric grid. As has been pointed out elsewhere in this book, a wind project doesn't provide power on an industry standard contract basis of 7 × 24 or 5 × 16 (e.g., five days per week, 16 hours per day). This requires that wind projects be backed up by fossil power plants or batteries. As stated in an October 25, 2010, White House briefing memo:

Renewables' intermittency problem limits the deployment of these technologies, which could be remedied by installing backup capacity (likely increasing the cost by 2 to 4 cts/kWh).

The memo later references DOE EIA, which shows that the cost of generating power from a wind project is 8.8 cts/kWh without subsidies and not taking backup power into account. This cost drops to 6.7 cts/kWh with the cash grant program and to 4.0 cts/kWh with the cash grant and the DOE loan guarantee program.

As stated by Nathan Myhrvold of Intellectual Ventures, and former chief technology officer of Microsoft:

Batteries suck. They're better than they used to be. They get a little better every year. The current rate of progress will take a very long time to get there. What it means is we have to roll up our sleeves and invest in something radically better.[3]

This raises the cost of wind power above that of a fossil power plant despite numerous claims that wind is "competitive" with fossil power plants. A fossil power plant can agree to a 5 × 16 contract, whereas a wind project can't meet this requirement due to the uncertainty of wind power

---

[2] Garrad Hassan, one of the world's leading wind energy consultants.
[3] Alan Murray, "The Next Smart Thing," *Wall Street Journal*, March 7, 2011.

production and the lack of battery technology. Power contracts also have penalties for nonperformance.

WIND PROJECT ECONOMICS

Most of the wind projects on the East Coast of the United States have a capacity factor in range of 30 percent. The wind projects that are offshore and in the plains operate in the high 30s to low 40s. A developer of a wind project is effectively purchasing an expensive machine that operates for only a relatively low number of hours. It also tends to produce power during times when the price for power is lowest. It is not uncommon that a wind project will produce most of its power during the early morning hours and during the winter season when the demand for power is lowest.

Wind developers seem to miss the use of a simple relative value calculation when evaluating the economics of their projects. It is frequently possible to buy the first or second lien of a distressed natural gas power plant that will provide a higher return than the equity of a wind project. There is also no development risk in the purchase of an existing debt issue. Wind developers that have purchased a number of wind turbines might agree to develop a project to reduce their holding cost. Even if the returns for these projects are low it might still make sense to develop and finance the project in order to reduce holding costs.

Unlike a fossil power plant, there is a limited amount of ongoing optimization possible with a wind plant. With a gas turbine combined-cycle project, there are a number of ways to reduce the overall variable cost of the project. This could include reductions in natural gas transportation expense, gas turbine engine upgrades, and reductions in variable operations and maintenance expense. First, cost is the key issue with a wind project.

Wind Project Tax Attributes

The economics of a wind project is based on the sale of energy, renewable energy credits, capacity payments (depending on the power market) and tax benefits. Regulated electric utilities are also able to place wind projects in rate base if they can obtain approval from their local public service commission. These entities don't have to worry about hedging the output of a power plant since they have a regulated rate base and only have to demonstrate that a particular power plant is "used and useful." As a result of this rate base, regulated utilities produce a lot of cash and have the appetite for the tax attributes that a wind project offers. They are a natural investor in their own and in third-party renewable power projects. Some utilities and their unregulated subsidiaries have recently invested in so many projects that they have actually hit the alternative minimum tax (AMT).

Due to the generous feed-in tariff program throughout the European continent, a large number of renewable power projects have been developed. Foreign wind developers have an extensive amount of experience from European power markets. However, they can have trouble using tax benefits due to their limited amount of U.S. source income. These investors will also be forced to look for tax equity investors after the cash grant expires at the end of 2011.

The tax attributes that are available to a wind project include the production tax credit (PTC) and, more recently, the investment tax credit (ITC). The ITC was first introduced by President Kennedy in 1962, and the program ran until 1969. At the time, the purpose of the ITC was to encourage the building of new infrastructure. In the past, wind projects qualified for only the PTC and not the ITC. The PTC can't pass through a lease, and as a result, wind projects could not use a leveraged leasing structure. The PTC is calculated based on mWh produced by the plant. Assuming a 100-mW wind plant, the PTC would be calculated as follows:

$$100 \text{ mW} \times 8,760 \text{ hrs/year} \times 30 \text{ percent availability}$$
$$\times \$22 \text{ mWh } (\$2011 \text{ PTC})$$
$$= \$5,781,600$$

The PTC is paid every year for a 10-year period and is indexed each year by inflation. As the prior calculation shows, an individual or a corporation would have to have a minimum annual tax burden of at least $5,781,600. The actual tax appetite would have to be higher since a wind project also produces a large amount of tax depreciation over the first five years of its life. Only a very wealthy individual would have this large a tax appetite.

Investment Tax Credit versus Production Tax Credit

It is very easy to run into passive income issues, and a developer without a tax appetite is forced to look only to widely held C corporations (as defined by the IRS) as his tax equity. Once the project starts operation, the PTC will continue to be paid as long as it produces electricity. In the past, the U.S. government has let the PTC expire every so often, which has led to boom-bust cycles in the wind industry. As of the time of this writing, the placed-in-service date for new wind power plants is December 31, 2012. The requirement to operate has led some merchant wind developers to bid their energy at $0/mWh so that they will be selected to run. This ensures that they will at least receive some tax benefits from the PTC.

The second tax benefit that a wind project receives is tax depreciation based on a five-year modified accelerated cost recovery schedule (MACRS). As a comparison, most fossil power plant equipment is depreciated over a 20-year period. According to Keith Martin in *Fortnightly's Green Utility*:

> *The accelerated depreciation on most renewables is worth about 26 cents per dollar of capital cost in terms of the value the tax savings one gets from claiming it. The tax credit varies in value, but for most projects it's a minimum of 30 cents per dollar capital cost.*[4]

In recent wind projects, developers have selected the cash grant and plan on using the net operating losses created by the five-year MACRS over the life of the project. Developers have found this to be a more efficient way to use these benefits then to bring in a tax equity investor.

Wind developers have been allowed to select the ITC, which is based on 30 percent of the project's capital cost. For example, if a project cost $100, then the developer would either receive $30 as a tax credit or, if the project started construction before the end of 2011, a check from the government for $30. This works especially well if a particular wind plant has a low dispatch factor. A higher dispatch factor might direct a developer to find tax equity that can use the large number of PTCs that would be produced by such a project. The cash grant has been especially popular with private equity funds. The limited partners of these funds are not taxpayers, and as a result, a PTC or ITC has no value to them. After the expiration of the cash grant, they will also be forced to go to the tax equity market. As of 2008, regulated utilities can now also qualify for the ITC. This has resulted in regulated utilities developing and siting solar projects in their own service territory and placing them in rate base.

WIND PROJECT POWER CONTRACTING

Wind projects that are developed by independent power producers frequently sign long-term power contracts with regulated utilities. These long-term contracts allow for a large amount of project finance debt. Utilities typically purchase the energy, capacity (at a discount to generator name plate), and renewable energy credits (RECs). Utilities typically don't like to purchase power from independent generators under long-term contracts

---

[4] Keith Martin, *Fortnightly's Green Utility*, March 1, 2011.

since they can't earn a return on these investments. The rating agencies can also treat these contracts as imputed debt. In the case of a wind project, they agree to the purchase since they are mandated by the state they operate in to purchase RECs. One megawatt hour of power produced by a renewable power plant is equivalent to one REC. By way of example, a 100-mW wind project that is operating at 30 percent availability would produce 100 mW × 8,760 hours/year × 30 percent availability = 262,800 mWh and 262,800 RECs.

There is currently no national standard for renewable credits and many states have already met their REC requirements. As a result, the prices for RECs have been depressed. Typically wind, solar, and other renewable energy projects have different REC categories. A merchant wind generator can have a difficult time determining fair value for RECs and obtaining a long-term contract for the RECs since they are not widely traded. Energy trading desks that quote energy pricing and enter into energy hedges don't quote RECs due to an overall lack of liquidity. Most analysts feel that REC programs would disappear if a carbon dioxide (CO_2) tax were implemented in the future. The reasoning is that prices for power would increase for fossil power plants that had to purchase CO_2 allowances. This overall pricing structure allows individual states to pay for a portion of the additional cost from renewables with RECs and the federal government to pay with tax credits.

Transmission Constraints

Wind projects are typically located far from population centers. This requires that transmission lines be developed to move their electricity to market. There have been cases in Texas that, during certain times of the day, a wind project is not able to sell its power due to transmission constraints. In fact, projects in West Texas have experienced 10 to 30 percent curtailment. Texas experienced low wind years in 2005 and 2007.[5] Both Texas and California are in the process of expanding their transmission infrastructure for renewable power projects. As mentioned earlier, the Federal Energy Regulatory Commission (FERC) doesn't have eminent domain authority to site transmission lines. It is not uncommon for transmission lines to take years to develop due to state-level oversight. There is also the issue of who pays for the transmission line. A wind project's economics might not be able to stand the additional cost of a transmission line.

Transmission issues can also come into play with existing power plants. Some coal plants have been forced to back down when a wind project is

---

[5] Alan Murray, "The Next Smart Thing."

able to produce power. This can result in a net increase in emissions since the coal plant will produce more emissions while operating at part load. The solution to this problem is a transmission upgrade that would direct the wind and/or coal plant's output to another part of the grid, resulting in a reduction of forced curtailment. As pointed out earlier in this book, this unscheduled curtailment by a coal facility is another example of the lack of overall coordination in the energy market. Operating power plants are still required for overall grid stability and can't be replaced instantly like a new web site.

Opportunities with Distressed Wind Projects

Wind turbine projects can also face environmental issues. The Beach Ridge Wind Farm in West Virginia was forced to reduce its size from 122 1.5-mW GE wind turbines or 183 mW to 67 1.5-mW wind turbines or 100.5 mW. This was due to a concern that the project could kill the Indiana bat, which is protected by the Endangered Species Act. The project also had to agree to restrictions on operating hours to protect the bats. The turbines are allowed to be in 24-hour operation between mid-November and April 1, when the bats are hibernating. For the remaining 7.5 months, the turbines may operate only in the daylight hours. The overall reduction in project size and restriction in operating hours make this an interesting candidate to watch for a potential future distressed investing opportunity.

A large number of older vintage wind projects were based on a 100 percent equity capital structure and are owned by deep-pocketed developers. These developers also tend to have a tax appetite and can be content with a low return since they are also enjoying both PTC and five-year MACRS depreciation. Newer vintage wind turbines have selected the cash grant and also have project finance debt. These projects are worth watching as potential distressed restructuring candidates. Since wind projects produce a small amount of EBITDA, overstating energy production or understating maintenance can easily cause a project to become distressed, leading to a possible debt default. There is also limited operating history and actual operating and maintenance cost for newer turbines. Wind turbine availability has been an issue. This situation occurs when the wind blows but the wind turbine is not available to make power. U.S. average wind turbine availability has been 93 percent as opposed to an expected 97 percent. Older wind energy production forecasts suffered from overoptimism. There was measurement bias and wind flow modeling issues for pre-2001 to 2003 wind projects. However, older wind projects mostly had 100 percent equity capital structures.

Newer wind forecasts have improved their accuracy, and newer projects have more leverage, which could expose them to future defaults.

Distressed wind opportunities could also include projects with a current power price calculated by avoided cost. This type of opportunity could be found with a renewable project that started operations with a front-loaded defined power price and then converted to avoided cost at a later date. In the current low-power price environment, a project of this type could easily shift into default. A study funded by the German government found that the availability of 10-year debt for wind parks had fallen by as much as 40 percent, while the cost of debt relative to benchmark interest rates more than tripled in 2009. The credit squeeze has eased in America, but less so in Europe. Wind farmers in Spain and Portugal find it hard to borrow because of worries that their governments may default or trim subsidies.[6]

WIND ENERGY PREDICTION

One of the challenges for wind projects comes from the projection of future production of electricity. Future megawatt hour production is calculated based on both on site and windspeed data from local sources such as an airport. One wind project investor stated that on its 54 wind farms, based on 90 years' worth of data, they had been underperforming by 10 to 11 percent. This is due to less wind turbine availability, wake effect due to terrain layout, and errors in the ultimate calculation from gross to net electric output. A drop in an availability of 10 percent would result in a drop in a typical wind project internal rate of return of 300 basis points. As wind turbines got taller, correlation with lower height meteorological towers produced bad data. Some developers have invested in met towers that are the same height as the actual wind turbines' hubs.

In Germany, for the first quarter of 2010 the Breeze Two wind project recorded wind conditions at 81 percent of the long-term average. The project's financial structure was based on more optimistic wind forecasts. As a result, the Class B debt of Breeze Two has been downgraded to C by Standard and Poor's, with a negative outlook.[7]

The main tool for assessing the wind resources at prospective wind farm sites is "measure, correlate, predict" (MCP) analysis. This involves

---

[6] "Green Energy: Wild Is the Wind," *Economist*, September 23, 2010.
[7] Standard & Poor's, "Summary: CRC Breeze Finance S.A.," www.alacrastore .com/research/s-and-p-credit-research-Summary_CRC_Breeze_Finance_S_A-916421 (accessed December 9, 2011).

comparing a short-term sample wind measurement at the site itself with years' worth of historical wind data taken from a nearby airport or permanent weather station. Using statistical models to take account of the difference between the target site and the reference site, developers can then build a detailed picture of the potential wind resources.[8]

Prediction of wind turbine output has been especially unreliable in complex terrain areas. These are locations with forests and mountains. To improve wind generation predictions, some companies are turning to laser- and sonar-based measurement instruments to complement MCP. In Denmark, which already gets nearly 20 percent of its electricity from wind power, a change in wind speed of one meter per second can translate into a change of 450 mW in national power output.[9]

Lenders to wind projects consider a P99, P90, and a P50 output. These outputs each have different percent chances, as follows:

- For a P50 output, there is a 50 percent chance that the wind project output will be higher than this value and a 50 percent chance it will be lower than this value.
- For a P90 output, there is a 90 percent chance that the wind project output will be higher than this value and a 10 percent chance it will be lower than this value.
- For a P99 output, there is a 99 percent chance that the wind project output will be higher than this value and a 1 percent chance it will be lower than this value.

Lenders also use a different debt coverage ratio depending on the P level used.

The focus for new wind turbines is to reduce weight by using carbon-fiber blade materials. Turbine manufacturers are also taking out their gearbox and using gearless turbines. In this case, the generator turns at the same speed as the rotor. The plant then uses an inverter to convert from direct current (DC) to alternating current (AC). There has also been a focus on remote sensing for performance anticipation and monitoring/trending for condition-based maintenance. Offshore wind projects face hard operating conditions, and these upgrades would also be required for their successful long-term operation.

---

[8] "Wind Forecasting: And Now, the Electricity Forecast," *Economist*, June 10, 2010.
[9] "Wind Forecasting," *Economist*.

SUMMARY

Wind projects have a low capacity factor and are dependent on tax benefits. Like solar power projects, wind projects face tough competition from abundant supplies of natural gas. The lack of a U.S. CO_2 tax and the challenge of siting new transmission lines also make it difficult for wind to compete with fossil power plants.

Chapter 10 further addresses electric power transmission constraints.

Electric Power Transmission

[T]he electrification of the whole country.
                                            —Vladimir Ilyich Lenin

Hundreds of wind farms and solar farms haven't been financed in the United States because they can't obtain electric power transmission access at a competitive cost and they cannot supply energy 24 hours a day, 7 days a week, 365 days a year. In short, these two major types of renewable power are available only intermittently each year—only one-third to one-sixth of the 8,760 hours in a year—and they are located too far from electricity grids or powerful transmission lines. Because of this immutable fact, it is very frequently extraordinarily difficult to sell debt in a wind power or solar renewable power project at an acceptable risk-adjusted rate of return.

OVERVIEW

Because key locations in the United States with the highest natural wind velocity or highest solar power available are usually a very long distance from major population centers or from major industrial or commercial centers, it is essential to invent new energy storage devices or new energy transmission technology to exploit vast amounts of the world's energy that is currently totally lost.

Likewise, unless power plants are located directly next to the power grid system, even when the wind blows furiously or the sun shines for hours, the excess energy will simply be "spilt, wasted, or "dissipated" because it cannot be used in our electric power grid system.

Major energy storage facilities and major energy transmission lines require huge financial investments. Very few companies, governments, independent system operators (e.g., PJM), or electric utilities in the United States have been willing to sign long-term energy supply contracts with power generation plants. Yet without major electric storage facilities and new transmission lines, few renewable power plant developers will take the huge financial risk of building brand new power plants. These plants will be vitally needed if our nation is to diversify away from fossil fuels such as coal, oil, or gas.

High-voltage electric transmission (or electric power transmission) is quite simply the bulk transfer of electrical energy from generating power plants to substations located near population centers. This high-voltage electric transmission is different from the local wiring between high-voltage substations and customers, which is usually referred to as *electric distribution*. Transmission lines, when interconnected with each other become high-voltage grid, or the National Grid in Great Britain. We have three grids in North America: the Eastern Interconnection, the Western Interconnection, and the Electric Reliability Council of Texas (ERCOT) Grid.

Transmission lines and distribution lines, which used to be owned by the same company, have recently been separated by regulatory reform into electricity transmission from the distribution business.

In the United States and Europe, transmission networks are called power grids or "The Grid." The regional transmission lines use three-phase alternating current (AC), although single-phase AC is sometimes used in railway electrification systems. High-voltage direct current (HVDC) technology is used only for very long distances (typically greater than 400 miles or 600 kilometers); submarine power cables, which are typically 30 miles to 50 km); or for connecting two AC networks that are not synchronized.

In the United States, electricity is transmitted at 110 volts or more to reduce energy lost in long distance. In the United Kingdom and most of Europe, electricity is transmitted in 220 volts (much higher voltage and requires far more security of grounding the current to prevent fires or human physical risks). In the United States, transmission power is usually transmitted through overhead power lines because it is cheaper than underground power cables. Voltage over 250 kilovolts (kV) is used to reduce energy lost in long distance transmission. Underground power cables are typically used in city areas or technically sensitive areas.

Because electrical energy cannot be stored, it must be generated in an amount equal to the actual power demanded. Electricity can be stored only in very limited amounts and inefficiently in batteries. For example, the battery of the Nissan Leaf can hold only approximately 24 kWh. To put this in context, a 100 mW wind turbine would generate at approximately 30 percent availability over the 8,760 hours in one year or 100 mW × 8,760 hours/year × 30 percent availability = 262,800 mWh/year or 262,800,000 kWh/year. This wind turbine project alone could supply 262,800,000/24 = 10,950,000 Nissan Leafs!

When electricity supply does not equal demand, transmission equipment can shut down, which, in extreme cases, can result in major regional blackouts of the entire Northwest or Northeast. In Europe, whole nations such as Greece have suffered blackouts. To guard against this major risk, electric transmission networks are interconnected into regional, national, or continental wide networks to assure multiple redundant alternative routes for power to flow. This could occur due to weather, equipment failures, or political or military crises. A great deal of analysis is done by transmission firms to estimate the maximum reliable capacity of each line, which is less than its physical limit in order to be certain there is spare capacity should there be any major power failure in various parts of this network or in different networks.

In electrical systems, high-voltage overhead conductors are not covered by insulation. The conductor material is nearly always a lightweight aluminum alloy with different new materials used to increase current carrying capacity and improve resistance to bad weather conditions, lightning, high wind, or boiling heat or below-freezing temperatures that can cause power failures or motion of the physical line causing gallop or fluttering or causing rise and fall of frequency or oscillation.

Underground Transmission

Underground transmission consists of power cables used in densely populated urban areas or under rivers to protect vital infrastructure such as bridges or tunnels, or to drastically reduce the danger from chemical emissions into the atmosphere or avoid dangerous electromagnetic fields.

Underground electric transmission is several times more expensive than overhead power lines, and the life-cycle cost of an underground power cable is two to four times the overhead power lines ($10 versus $20 to $40 per foot). The cost of finding and repairing overhead wires is small and can take time measured in hours, whereas the time needed to repair underground cables can be measured in days or weeks.

As a result, extra (or redundant) cables are run. Because underground cables touch the earth, they cannot be maintained live, whereas overhead power lines can be. Not only do underground cables produce large charging currents, they make voltage control difficult. Nevertheless, in some cases, the advantages of underground cable outweigh its high costs.

The history of electric power direct current (DC) transmission truly took off in 1882. In 1886, in both Massachusetts and Italy, alternating current (AC) distribution systems were installed. In 1888, Nikola Tesla gave his lecture called "A New System of Alternating Current, Motors and Transformers," describing equipment that allowed efficient power generation and use of multi phase alternating current. The key was the transformer, and Tesla's polyphase and single-phase induction motors were essential for a combined AC distribution system for both lighting and machinery. This "universal system" used transformers to step up voltage from generators to high-voltage transmission lines, and then to step down voltage to local distribution circuits or industrial customers.

By correctly selecting utility frequency, both lighting and motor loads could be served. Rotary converters and later mercury-arc valves and other rectifier equipment allowed DC to be provided where needed. Generating stations and loads using different frequencies could be interconnected using rotary converters. By using common generating plants for every type of load, important economies of scale were achieved, lower overall capital investment was required, and load factor on each plant was increased, allowing for higher efficiency and a lower cost for the consumer, resulting in increased overall use of electric power.

By allowing multiple generating plants to be interconnected over a region, electricity production cost was reduced. The most efficient plants could be used to supply the varying loads during the day. Reliability was improved and capital investment cost was reduced, since stand-by generating capacity could be shared over many more customers and a wider geographic area. Remote and low-cost sources of energy, such as hydroelectric power or mine-mouth coal, could be exploited to lower energy production cost. The first transmission of three-phase AC using high voltage took place in 1891 during the international electricity exhibition in Frankfurt. A 25-kV transmission line, approximately 175 km long, connected Lauffen on the Neckar and Frankfurt.

Voltages used for electric power transmission increased throughout the twentieth century. By 1914, 55 transmission systems, each operating at more than 70 kV, were in service. The highest voltage then used was 150 kV. The rapid industrialization in the twentieth century made electrical transmission lines and grids a critical part of the infrastructure in most industrialized nations. Interconnection of local generation plants and small distribution networks was greatly spurred by the requirements of World War I, where large electrical generating plants were built by governments to provide power to munitions factories. Later, these plants were connected to supply civil loads through long-distance transmission.

Bulk Power Transmission

Today, a regional transmission network is managed on a regional basis by a transmission system operator. Transmission efficiency is enormously improved by many devices that increase the voltage and proportionately reduce the current in the conductors, thus keeping the power transmitted nearly equal to the power input. The reduced current flowing through the line reduces the losses in the conductors. According to Joule's Law, energy losses are directly proportional to the square of the current. Therefore, reducing the current by a factor of two will lower the energy lost to conductor resistance by a factor of four.

This change in voltage is usually achieved in AC circuits using a step-up transformer. DC systems require relatively costly conversion equipment, which may be economically justified for particular projects, but is less common currently.

A transmission grid is a network of power stations, transmission circuits, and substations. Energy is usually transmitted within a grid with three-phase AC. Single-phase AC is used only for distribution to end users since it is not usable for large polyphase induction motors. In the twentieth century, two-phase transmission was used but required either three wires with unequal currents or four wires. Higher-order phase systems require more than three wires but deliver marginal benefits.

The capital cost of electric power stations is so high, and electric demand is so variable, that it is often cheaper to import some portion of the needed power than to generate it locally. Because nearby loads are often correlated (hot weather in the Southwest portion of the United States might cause many people to use air conditioners), electricity often comes from distant sources. Because of the economics of load balancing, wide-area transmission grids now span across countries and even large portions of continents. The web of interconnections between power producers and consumers ensures that power can flow, even if a few links are inoperative.

The unvarying (or slowly varying over many hours) portion of the electric demand is known as the base load and is generally served best by large facilities (which are therefore efficient due to economies of scale) with low variable costs for fuel and operations. Such facilities might be nuclear or coal-fired power stations or hydroelectric power plants, while other renewable energy sources such as concentrated solar thermal and geothermal power has the potential to provide base load power. Renewable energy sources such as solar photovoltaics, wind, wave, and tidal are, because of their intermittency, not considered base load, but they can still add power to the grid. The remaining power demand, if any, is supplied by peaking power plants, which are typically smaller, faster-responding, and higher-cost sources, such as combined-cycle or combustion turbine plants fueled by natural gas. Peaking or simple-cycle gas turbine engines also have a relatively higher emissions profile than combined-cycle gas turbines.

Thus, distant suppliers can be cheaper than local sources. The Linden Variable Frequency Transformer connects the PJM grid with New York City Zone J. Multiple local sources (even if more expensive and infrequently used) can make the transmission grid more fault tolerant to weather and other disasters than can distant suppliers.

Long distance transmission allows remote renewable energy resources to be used to displace fossil fuel consumption. Hydro and wind sources can't be moved closer to populous cities, and solar costs are lowest in remote areas where local power needs are minimal. Connection costs alone can determine whether any particular renewable alternative is economically sensible. Costs can be prohibitive for transmission lines, but various proposals for massive infrastructure investment in high-capacity, very-long-distance super grid transmission networks could be recovered with modest usage fees. With current technology there are still limits on the ultimate distance that power can be moved.

GRID INPUT, LOSSES, AND EXIT

At the generating plants, energy is produced at a relatively low voltage, between about 2.3 kV and 30 kV, depending on the size of the unit. The generator terminal voltage is then stepped up by the power station transformer to a higher voltage (115 to 765 kV AC, varying by country) for transmission over long distance.

Transmitting electricity at high voltage reduces the fraction of energy lost to resistance. For a given amount of power, a higher voltage reduces the current and thus the resistive losses in the conductor. For example, raising the voltage by a factor of 10 reduces the current and therefore the I^2R losses by a factor of 100, provided the same-sized conductors are used in both cases. Even if the conductor size (cross-sectional area) is reduced 10-fold to match the lower current, the I^2R losses are still reduced 10-fold.

Transmission and distribution losses in the United States were estimated at 6.6 percent in 1997 and 6.5 percent in 2007. In general, losses are estimated from the discrepancy between energy produced (as reported by power plants) and energy sold to end customers; the difference between what is produced and what is consumed constitute transmission and distribution losses.

At the substations, transformers reduce the voltage to a lower level for distribution to commercial and residential users. This distribution is accomplished with a combination of subtransmission (x ke to 132 kV) and distribution (3.3 to 25 kV). Finally, at the point of use, the energy is transformed to low voltage (varying by country and customer requirements).

HIGH-VOLTAGE DIRECT CURRENT

HVDC is used to transmit large amounts of power over long distances or for interconnections between asynchronous grids. When electrical energy is required to be transmitted over very long distances, it is more economical to transmit using direct current instead of alternating current. For a long transmission line the lower losses and reduced construction cost of a DC line can offset the additional cost of converter stations at each end.

HVDC is also used for long submarine cables because over about 30 km length, AC can no longer be applied. In that case, special high-voltage cables for DC are built. Many submarine cable connections up to 600 km length are in use nowadays. HVDC links are sometimes used to stabilize against control problems with the AC electricity flow.

The amount of power that can be sent over a transmission line is limited. The origins of the limits vary depending on the length of the line. For a short line, the heating of conductors due to line losses sets a thermal limit. If too much current is drawn, conductors may sag too close to the ground, or conductors and equipment may be damaged by overheating. For intermediate-length lines on the order of 100 km (62 mi), the limit is set by the voltage drop in the line. For longer AC lines, system stability sets the limit to the power that can be transferred.

Up to now, it has been almost impossible to foresee the temperature distribution along the cable route, so that the maximum applicable current load was usually set as a compromise between understanding of operation conditions and risk minimization. The availability of industrial distributed temperature sensing (DTS), which measures temperatures along the cable in real time, is a first step in monitoring the transmission system capacity. This monitoring solution is based on using passive optical fibers as temperature sensors, either integrated directly inside a high-voltage cable or mounted externally on the cable insulation.

CONTROLLING THE COMPONENTS OF THE TRANSMISSION SYSTEM

To ensure safe and predictable operation, the components of the transmission system are controlled with generators, switches, circuit breakers, and loads. The voltage, power, frequency, load factor, and reliability capabilities of the transmission system are designed to provide cost-effective performance for the customers.

Load Balancing

The transmission system provides for base load and peak load capability, with safety and fault tolerance margins. The peak load times vary by region largely due to the industry mix. In very hot and very cold climates, home air conditioning and heating loads have an effect on the overall load. They are typically highest in the late afternoon in the hottest part of the year and in midmorning and midevenings in the coldest part of the year. This makes the power requirements vary by the season and the time of day. Distribution system designs always take the base load and the peak load into consideration.

The transmission system usually does not have a large buffering capability to match the loads with the generation. Thus, generation has to be kept matched to the load, to prevent overloading failures of the generation equipment.

Multiple sources and loads can be connected to the transmission system, and they must be controlled to provide orderly transfer of power. In centralized power generation, only local control of generation is necessary, and it involves synchronization of the generation units, to prevent large transients and overload conditions.

Failure Protection

Under excess load conditions, the system can be designed to fail gracefully rather than all at once. Brownouts occur when the supply power drops below the demand. Blackouts occur when the supply fails completely. Rolling blackouts, or load shedding, are intentionally engineered electrical power outages, used to distribute insufficient power when the demand for electricity exceeds the supply.

ELECTRICITY MARKET REFORM: COSTS AND MERCHANT TRANSMISSION ARRANGEMENTS

Some regulators regard electric transmission to be a natural monopoly, and there are moves in many countries to separately regulate transmission. Spain was the first country to establish a regional transmission organization. In that country, transmission operations and market operations are controlled by separate companies. Spain's transmission system is interconnected with those of France, Portugal, and Morocco. In the United States and parts of Canada, electrical transmission companies can operate independently of generation and distribution companies.

The cost of high-voltage electricity transmission (as opposed to the costs of electricity distribution) is comparatively low, compared to all other costs are rising in a consumer's electricity bill. In the United Kingdom, transmission costs are about 0.2 p/kWh compared to a delivered domestic price of around 10 p/kWh.

Merchant transmission is an arrangement where a third party constructs and operates electric transmission lines through the franchise area of an unrelated utility. Advocates of merchant transmission claim that this will create competition to construct the most efficient and lowest-cost additions to the transmission grid. Merchant transmission projects typically involve DC lines because it is easier to limit flows to paying customers. The cost for a typical transmission line is $1,000 to $1,500/kW. Most of these merchant projects connect disparate grids, which require a conversion from AC to DC and then back to AC in order to synchronize between grids. For shorter transmission distances between different grids, a variable frequency transformer can be used. The FERC has been very supportive of these merchant transmission projects.

The challenge for merchant projects is locating them in areas that are environmentally sensitive. To avoid this problem, transmission lines are frequently placed under water. This can be a problem when the lines are buried in a sensitive area like a lake. Transmission projects can suffer from NIMBY (not in my backyard) and BANANA (build absolutely nothing anywhere near anybody).

Merchant transmission projects have been developed by holding an open-season bidding process. During the open season, potential transmission shippers bid for the amount and term of capacity they require. The successful transmission developer then selects one or more bidders and uses their purchase commitment to finance the project. There can be concerns in certain cases that a particular power market is moving its low-cost power to a high-cost market and providing the buyer with an unfair advantage. Some of these projects have been structured using a holdco/opco-type structure (*holdco* denotes holding company; *opco* denotes operating company). In this case, the FERC provides an incentive return on equity for the project and allows the project to obtain holdco leverage. This holdco leverage helps the project magnify its equity return.[1]

There is currently no federal eminent domain to support the siting of new transmission lines. In the case of natural gas pipelines, the FERC does have eminent domain to support siting. The potential closure of the Indian Point nuclear power plant in Westchester County, New York, could result in the development of a new merchant transmission line. This new HVDC

---

[1] Power assets are placed at the "opco." The FERC only looks at the amount of debt and equity at the opco and it is not concerned about the capital structure of "holdco."

merchant transmission line could move power from upstate New York, where there is excess capacity, to the Indian Point area. The current capacity market in most parts of the United States is only for a three-year period and can't support the building of a new merchant transmission line or a power plant. Al Coase, the eminent economist, would say this is not a market failure, only that the market needs to be further developed.

A U.S.-based transmission superhighway is an impractical way to deliver a large amount of power from renewable power plants. Onshore wind is also becoming difficult to site in some locations. It is possible that offshore wind could become competitive in the future if there is a relatively high CO_2 tax, natural gas prices increase, the United States becomes serious about meeting its future power needs from renewables, and the cost of offshore wind projects comes down. At the time of this writing, the cost of production from offshore wind is not competitive with other resources. Developing and permitting offshore wind projects is also difficult. The Cape Wind project spent $60 million to develop its yet unfinanced offshore wind project in Cape Cod, Massachusetts.

From a financing standpoint transmission projects can now use a master limited partnership (MLP) structure. An MLP is a publicly traded partnership. MLPs are restricted to firms that own natural gas pipelines and have restrictions on the amount of electric power assets that they can own. An MLP structure provides a firm with a tremendous tax advantage. Every firm would become an MLP if it could.

Operating merchant transmission projects in the United States include the Cross Sound Cable from Long Island, New York, to New Haven Connecticut; Neptune RTS Transmission Line from Sayreville, New Jersey, to New Bridge, New York; and the Linden VFT from Linden, New Jersey, to Staten Island, New York. The 660-mW Neptune line provides 20 percent of Long Island's electricity. The thesis for all of these projects is to move power from a low-cost, noncongested area to a high-cost, congested area. Except for Linden, each of these projects was financed based on a long-term offtake contract with a traditional electric utility. These transmission projects also provide electric power supply and fuel diversity in that power is moved from a mostly coal/nuclear region to a natural gas–based region.

ADDITIONAL CONCERNS

Effects on Health

The preponderance of evidence does not suggest that the low-power, low-frequency, electromagnetic radiation associated with household current constitutes a short- or long-term health hazard. Some studies have found statistical correlations between various diseases and living or working near power lines, but no health effects have been substantiated for people not living close to power lines.

There are established biological effects for acute high-level exposure to magnetic fields. In a residential setting, there is "limited evidence of carcinogenicity in humans. In particular, childhood leukemia is associated with average exposure to a residential power-frequency magnetic field above 0.3 to 0.4 UT. (Ultrasonic testing [UT] is a measurement of magnetic flux density which causes potential harm to humans.) These levels exceed average residential power-frequency magnetic fields in homes in North America.

Government Policy

In the United States, power generation is growing four times faster than transmission, but big transmission upgrades require the coordination of multiple states, a multitude of interlocking permits, and cooperation among a significant portion of the 500 companies that own the grid. Control of the grid is balkanized, and even former Energy Secretary Bill Richardson refers to it as a "third world grid." There have been efforts in the United States and the European Union to confront the problem. The U.S. national security interest in significantly growing transmission capacity drove passage of the 2005 energy act giving the Department of Energy (DOE) the authority to approve transmission if states refuse to act. However, soon after using its power to designate two national interest electric transmission corridors, 14 senators signed a letter stating the DOE was being too aggressive.

Superconducting Cables

High-temperature superconductors promise to revolutionize power distribution by providing lossless transmission of electrical power. The development of superconductors with transition temperatures higher than the boiling point of liquid nitrogen has made the concept of superconducting power lines commercially feasible, at least for high-load applications. It has been estimated that the waste would be halved using this method, since the necessary refrigeration equipment would consume about half the power saved by the elimination of the majority of resistive losses.

Some companies such as Consolidated Edison and American Superconductor have already begun commercial production of such systems. In one hypothetical future system called SuperGrid, the cost of cooling would be eliminated by coupling the transmission line with a liquid hydrogen pipe-line. Superconducting cables are particularly suited to high-load-density areas such as the business districts of large cities, where purchase of an easement for cables would be very costly.

Single-wire earth return (SWER) or single-wire ground return is a single-wire transmission for supplying single-phase electrical power for an electrical grid to remote areas at low cost. It is principally used for rural electrification, but also finds use for larger isolated loads such as water pumps and light rail. SWER is also used for HVDC over submarine power cables.

Both Nikola Tesla and Hidetsugu Yagi attempted to devise systems for large-scale wireless power transmission, with no commercial success. Wire-less power transmission has been studied for transmission of power from solar power satellites to earth. A high-power array of microwave transmit-ters would beam power to a rectenna. Major engineering and economic challenges face any solar-power satellite project.

Security of Control Systems

The federal government of the United States admits the power grid is sus-ceptible to cyber-warfare. The U.S. Department of Homeland Security works with industry to identify vulnerabilities and to help industry enhance the security of control system networks. The federal government is also working to ensure that security is built in as the United States develops the next generation of "smart-grid" networks.

SUMMARY

Renewable projects are often located in remote areas and are dependent on transmission to serve load. Huge financial expenditures on new electric transmission lines or underground transmission cables or new wireless transmission, or satellite transmission, will need to be researched, eval-uated, and deployed carefully in the next 5 to 10 years, in order for renew-able energies (such as wind, solar PV, or solar thermal energy; tidal and wave energy; etc.) to have any significant opportunity to drastically reduce the world's reliance on fossil fuels.

In addition, the U.S. federal government admits that the power grid is susceptible to cyber-warfare, so there will be costs associated with improv-ing things on this front, too. The U.S. Department of Homeland Security works with the industry to identify vulnerabilities and to help the industry enhance the security of control system networks. The federal government is also working to ensure that security is built in as the United States develops the next generation of "smart-grid" networks.

Enormous advances in electric energy transmission have been made in the past 140 years. It is essential that equal if not greater investments in advances in innovative electrical transmission systems plus innovative electric power storage technologies will have to be fully researched, developed, and launched into mass production on a global scale if the range of renewable energies is to succeed in creating a drastic decline in toxic emissions, reduction in global warming, and transforming the world to a decline in pollution and global long-term self-sufficiency in clean energy.

In Chapter 11, we discuss natural gas power plants.

Natural Gas Power Plants

Is Poker a game of chance? Not the way I play it.

—W.C. Fields

Power price forecasts in the United States are based on the assumption that all future fossil power plants will be natural gas turbine engine fired. The revenue requirement for a natural gas turbine fired power plant has become the benchmark or proxy unit that renewable and other fossil power plants are compared against. It is currently next to impossible to obtain an air permit for a new coal plant in the United States. In addition, the low price of natural gas and the high price of coal have made it difficult for coal plants to compete. Improvements in gas turbine engine heat rate have made competing against them even more difficult. Operating coal plants face new regulation on sulfur dioxide (SO_2), nitrous oxide (NOx), particulate, and mercury, which might make smaller projects uneconomic. Natural gas represents approximately 22 percent of electricity generation.[1]

GAS TURBINE ENGINES

Gas turbine engines have become the technology of choice for the generation of electric power. Over the past 20 years, there have been tremendous advances in gas turbine engine technology. This has resulted in improved part-load performance, increased power output, reduction in heat rate, and large decreases in emissions. Gas turbine engines are frequently used in a combination with one or more steam turbines to create a combined-cycle power plant. The term *combined cycle* comes from the fact that the gas

---

[1] Natural Resources Defense Council, October 2008.

turbine engine is based on the Brayton Cycle, and the steam turbine is based on the Rankine Cycle. The exhaust heat from the gas turbine engine is captured in a heat recovery steam generator (HRSG), which produces steam to drive a steam turbine. Additional steam can be generated in the HRSG by burning additional natural gas in a duct burner. This additional steam can be used to provide additional power and/or process steam. Gas turbine engines can be configured on a one-on-one basis, which is one gas turbine and one steam turbine, or a two- or three-on-one configuration, which refers to two or three gas turbines on one steam turbine.

Unlike coal-fired power plants, natural gas–fired power plants can obtain an air permit and local approval relatively quickly. In the United States, there is an extensive infrastructure of natural gas pipelines and natural gas storage. It is a lot easier dealing with natural gas transportation and commodity than with the supply, handling, and resulting ash disposal of a coal power plant. There is no concern with the disposal of ash or any future ash disposal regulations issue with a natural gas–fired power plant.

The best greenfield sites for new gas turbine engines are at the intersection of one or more natural gas pipelines and one or more electric substations. Frequently, a gas turbine power plant will seek to have access to two or more natural gas transmission lines for reliability purposes. A natural gas transmission line that supplies natural gas at a high pressure removes the need for the power plant to purchase a natural gas compressor. A natural gas compressor increases the parasitic load at the power plant, resulting in a decrease of power output and an increase in heat rate. Gas turbine power plants will also attempt to get language in their air permits that allows them to operate for up to 30 or more days on fuel oil or another backup fuel. There may be certain times of the year when it makes economic sense to run on fuel oil instead of natural gas or during times when natural gas commodity or transportation is interrupted. Natural gas projects can also increase their cash flow by reselling some of their firm transportation.

As older coal plants are shut down, these brownfield sites could be repowered with new natural gas turbine–based projects. The objective will be to use the emissions profile from the old coal-fired power plant in order to "net out" the emissions from the new gas-fired power plant. Gas turbine combined-cycle (GTCC) power plants require water for their steam turbine cycle and may locate next to a wastewater treatment plant in order to use the "gray" water produced by the treatment facility. Makeup water would still have to come from traditional sources. Typically, areas that host wastewater treatment plants are also located in industrial areas that will support the development of a new power plant. U.S. states have been supportive of locating new combined-cycle power plants on existing brownfield industrial sites. All of these issues make it easier to obtain local support for the project.

BENEFITS OF GAS TURBINE ENGINES

Unlike coal and nuclear power plants, newer combined-cycle plants are able to operate at part-load and still meet low emission levels. Combined-cycle plants are also able to operate only the gas turbine engine portion of the power plant. Natural gas plants can provide backup power to the intermittent performance of wind and solar power plants. It is important for developers to consider "optionality" issues in their air permit. A gas turbine engine with quick start capability can capture the volatility that will be present in future power prices due to the intermittency of renewable power plants. This allows for the possibility of 24-hour-a-day, seven-day-a-week operation and the ability to start the gas turbine engine quickly without violating any air permit conditions. If the price for power is high, the air permit has to allow the generator to dispatch in order to capture this price.

It is important for a gas turbine engine power plant to be able to ride through periods of low-priced power without being forced to turn down to zero output. Starting a gas turbine engine from a cold start is expensive, takes time, and might cause the project to miss a high price for power. Gas turbine engine manufacturers are offering more products that improve the part-load performance and ramp rate of their new and existing engines. These upgrades can also help the gas turbine engine meet its air permit constraints. As previously explained, coal and nuclear plants are inefficient at part-load performance.

GAS TURBINES AND CO$_2$

When and if there is a tax on carbon natural gas, combined-cycle plants have an advantage over coal plants in that their carbon emission stream is approximately 50 percent less than a coal plant. There are currently some studies under way to calculate the methane and other greenhouse gas emissions stream that also result from the production of shale gas. This incremental stream of emissions would also be added to the emission stream from the operating gas turbine engine plant. Initial calculations show that even taking into account natural gas production from shale, combined-cycle power plants have the lowest carbon emissions for a fossil power plant.

It is also possible to shift out the hydrogen and the CO$_2$ from natural gas by using a shift reactor. The gas turbine engine would operate on 100 percent hydrogen, and the CO$_2$ stream could be buried underground. In addition to the burial challenge, gas turbine engines have only limited history on operation with 100 percent hydrogen, and it might not be possible to obtain a guarantee on performance from the engine manufacturer.

Gas turbine engines that use dry low NOx (DLN) combustors also have limits on the amount of hydrogen in the fuel that they consume. The latest performance standard is a maximum of only 5 percent hydrogen. DLN has become the standard technology for new gas turbine engines and a large number of the operating gas turbine engines. The lack of these guarantees would make it difficult to obtain long-term project financing on a non-recourse basis.

Cogeneration

Cogeneration is the sale of steam from gas turbine engine power plants to large, industrial users of steam. Cogeneration can also be defined as working one fuel twice. Cogeneration captures the heat that would normally be wasted while generating power and supplying the heating and/or cooling needs of the user.[2]

It is possible for the industrial user to shut down its operating boiler and to purchase all of its steam needs from the gas turbine power plant. This approach also creates emission reduction, which can help the power plant obtain its air permit. In a nonattainment location or an area that exceeds federal standards for a particular pollutant, this may be the only way to obtain an air permit. As described earlier, a GTCC project will typically produce more electricity than steam. With a coal boiler steam turbine configuration, the reverse is true.

The quality of the steam (e.g., steam temperature and pressure) will determine the ultimate plant heat rate and fuel chargeable to power. This calculation takes into account the amount of British Thermal Units (Btus) sold as steam and adjusts the heat rate down accordingly. A net reduction in CO_2 and other pollutants such as NOx and SO_2 results from an efficient cogeneration process. It may also be possible for the gas turbine to burn some of the waste fuel provided by the industrial host. This can allow the power plant to reduce its strike price as compared to its competitors.

GAS TURBINE OPERATIONS

Gas turbine projects typically execute a long-term service agreement (LTSA) with the gas turbine engine supplier that provides for maintenance support on the gas turbine engines. Under the LTSA, the engine supplier will supply all parts and labor for the scheduled and unscheduled maintenance of the generators and ancillaries during a term of typically up to 15 years from

---

[2] www1.eere.energy.gov/industry/distributedenergy/pdfs/chp_report_12-08.pdf.

execution of a contract of sale for the gas turbines. Based on the anticipated mode of operation for the facility, the term is expected to be 10 years from the start of commercial operation and will include a hot gas pass inspection and two combustion inspections. The engine manufacturer's fee for the scheduled maintenance service is based on a lump sum for the term to be paid in proportion to the starts accrued on the gas turbines. The LTSA typically includes a provision for the engine manufacturer to remotely monitor the condition of the engines to facilitate data acquisition and detect abnormalities early with the intent of maximizing engine reliability and plant availability.

It is possible to burn landfill gas in a gas turbine engine. The gas turbine engine usually requires that the landfill gas be cleaned to a level that could make the project uneconomical. Reciprocating engines from firms such as Caterpillar tend to be the best choice to burn landfill gas. Typical reciprocating engine size tends to be 1 megawatt (mW). It is rare to see a landfill gas project that is larger than 5 mW in size. Landfill gas consists of mostly methane, which is a greenhouse gas that is 21 times stronger than CO_2. As a result, there is actually a reduction in overall greenhouse gas emissions from the combustion of landfill gas. Along with cogeneration, this is a technology that is available today that can economically and meaningfully reduce CO_2 emissions. Due to this fact, landfill gas projects can qualify for renewable energy credits (RECs). It was pointed out in other chapters that investor-owned utilities will agree to purchase energy from projects of this type in order to meet their REC requirements.

The local natural gas grid can contain impurities, including ethane from natural gas liquid production and industrial waste gases. These impurities will have to be removed prior to the large-scale implementation of compressed natural gas vehicles. In some cases, these gases can be cleanly burned directly in a gas turbine engine, resulting in a net reduction in emissions.

SUMMARY

Gas turbine engine technology continues to improve. This results in lower heat rates, improved emissions, and increased power output. This, along with the low price of natural gas, no material CO_2 tax, and no technology to remove and sequester CO_2 at scale, makes it difficult for renewable projects to compete against natural gas turbine engine power plants.

Chapter 12 reviews coal-fired power plants, another existing technology that renewables have to compete against.

Coal-Fired Power Plants

The numbers should be talking to you.
                    —Larry Grundmann, energy expert

Why would a book on renewable energy discuss coal-fired plants, much less devote an entire chapter to them? Renewable power plants don't operate in a world without other power plants. Renewable power projects can't supply power on a 7×24 basis, and each grid has to rely on fossil fuel–based power plants. As previously discussed, there is no battery technology that can store power produced by renewable power plants during off-peak times. Coal-fired power plants create the following discharges, which require environmental consideration:

- The discharge of particulate and gaseous emissions.
- The discharge of heat or thermal energy.
- The discharge of solid and liquid wastes.
- Noise.

Only a few countries have access to a large amount of hydro and geothermal power projects with a high enough availability to achieve grid stability. Understanding how coal plants operate and their economics is critical for renewable power investors.

COAL'S HIGH OUTPUT CAPACITY

Coal power plants produce approximately 50 percent of the energy in the United States. Supplies of high heat content coal are found in large, mineable quantities throughout the United States. The coal-to-cover ratio in the United States as compared to other countries is relatively low. The heat content or British Thermal Unit (Btu)/lb content of U.S. coal also tends to be higher than that found in other countries. One of us has worked on coal-fired power projects in other countries that would require a large subsidy to make electric power competitively. This is due to the fact that the coal was buried deep underground and contained a low Btu content. It is important for new investors to understand that the $/ton price for coal or biomass is not as important at the $/MMBtu price. If one pays $20/ton for an 8,000 Btu/lb coal, this is equivalent to $20/ton × 1 ton/2,000 lb × lb/8,000 Btu 1,000,000 Btu/MMBtu = $1.25/MMBtu. Power plant developers will often make this mistake when they consider a fuel that is cheaper in price on a $/ton basis but expensive on a $/MMBtu. In order to determine a delivered price for coal, one would add in the cost of trucking and/or rail to the $20/ton in order to determine the delivered cost of coal on a $/MMBtu basis. This is especially true when one considers the use of wood waste and other opportunity fuels.

A natural gas–fired power plant produces approximately half of the carbon dioxide (CO_2) emissions of a coal-fired plant. The following is the calculation of CO_2 production from a new 500-megawatt (mW) coal plant with a 9,000 Btu/kWh heat rate operating at 90 percent availability:

$$500,000 \, \text{kWh/hr} \times 9,000 \, \text{Btu/kWh} \times \text{MMBtu}/1,000,000 \, \text{Btu}$$
$$= 4,500 \, \text{MMBtu/hr}$$

$$4,500 \, \text{MMBtu/hr} \times 206.7 \, \text{lb of } CO_2/\text{MMBtu} \times 1 \, \text{ton}/2,000 \, \text{lb}$$
$$\times 8,760 \, \text{hrs/yr} \times 90 \, \text{percent availability}$$
$$= 3,666,651 \, \text{tons/yr}$$

At an allowance cost of $10/ton of CO_2, this would be an additional yearly expense of $36,666,510/year.

Transportation of coal via rail and/or truck tends to be more developed in the United States than in other countries. At the time of this writing, there is no federal CO_2 tax or cap-and-trade program that would greatly hurt the economics of coal-fired power plants and conversely help renewable power plants. Most experts acknowledge that when the CO_2 comes about, it will be in the range of only $10/ton. It is difficult for the U.S. Environmental Protection Agency (EPA) to regulate CO_2 emissions since there is currently no economically proven, available technology to control CO_2 emissions. The EPA is very limited under best available control technology (BACT) regulations to restricting CO_2 emissions. Earlier in the book we discussed how controlling other pollutants such as sulfur dioxide (SO_2) and nitrous oxide (NOx) was easy due to proven and operating control technologies such as SO_2 scrubbers.

LIFE OF A COAL PLANT

Once a power plant is given its operating permits, it is difficult for the plant to be shut down by the EPA unless the proposed clean air transport rule (CATR) for SO_2 and NOx or maximum achievable control technology (MACT) for mercury is promulgated. As a comparison, a natural gas–fired combined-cycle plant would produce 50 percent less CO_2 emissions.

Coal plants are frequently located in areas where there are capacity/reliability requirements or transmission constraints or where there is a need for voltage or other types of grid support. In these cases, it would be difficult to force a coal-fired power plant into retirement. As stated elsewhere in this book, a typical wind plant operates only 30 percent of the time and will require backup from operating power plants. It is possible that a coal-fired power plant could be repowered by a natural gas–fired facility. This will require at least a one-year permitting period and an additional two- or three-year construction period. It will probably be necessary to upgrade the local natural gas pipeline system. Current market prices for power don't support the economics of a new build or a repowering with a natural gas combined-cycle power plant. There is a probability that natural gas prices will climb into the $5 or $6/MMBtu range in the future, which would help make coal plants more competitive. When all of these issues are considered, it might be possible that a larger number of coal plants would continue to operate in the future.

Some studies have come out that state that 35 to 50 gigawatts (gW) of operating coal plants under 300 mW would close down in the near future. Some lenders have also expressed concern about financing coal-fired plants in the future due to uncertainty about coal plants being able to achieve full cost recovery for CO_2 emissions. This misses the issue discussed above. A number of plants in this size range are well maintained and have also installed scrubbers or selective catalytic reduction (SCR) for NOx control and will remain in compliance. In the current power market, it will also be difficult to recover the full capital cost of building a new combined-cycle power plant. Once existing coal holding ponds are filled, it is possible that future ash disposal will require Class 1 status. A ruling of this type would drastically increase the ash disposal cost for a power plant.

There will also be a number of opportunities to convert coal plants below 70 mW in size to fire biomass. Biomass is typically defined as wood waste, construction waste, and forest trimmings. It is difficult to find a large enough supply of wood waste to supply a project larger than 70 mW in size. For example, a 50-mW biomass project would require 50,000 kWh/hr × 12,000 Btu/kWh × lb/4,500 Btu × ton/2,000 lb = 66.67 tons/hr or, based on 8,760 hours/year, 584,029 tons/year. Over a 30-year operating life, this is 17,520,876 tons of biomass!

The nameplate megawatt output of an existing coal plant that is converted to biomass will also drop by 20 percent or more. This is due to the fact that a typical biomass fuel has a heat content of only 4,500 Btu/lb, while a typical U.S. coal is over 10,000 Btu/lb. A coal-fired project that converted to biomass would also be eligible for renewable energy credits (RECs) and, depending on the amount of capital required for conversion, could qualify for tax-exempt debt or energy tax credits.

EXTENDING COAL PLANT OPERATIONS

The operation of coal-fired power plants can potentially be extended by selling steam to a nearby industrial host. The sale of steam would act as an additional revenue source and help in an overall reduction of CO_2 since the industrial host would be able to shut down its existing boilers. In a number of cases, these boilers tend to be old and relatively high emitters of air pollutants. This steam sale could cause operating issues for the coal plant when it was not called to dispatch by the local grid or power purchaser. A number of older plants are currently facing an issue that their steam sales were priced too low. This can especially be a problem when an industrial customer requires steam with both a high temperature and pressure. Boiler steam turbine–based cogeneration systems work best with steam hosts that require a large amount of process steam. Put another way, the thermal/power ratio is higher for boiler steam turbine as opposed to gas turbine engine combined-cycle projects. A new boiler can also be sized to meet a required steam and power need, while gas turbine engines are produced in standard sizes.

Operating circulating fluidized bed (CFB) boilers will also have interesting future operating potential. These facilities already have the ability to control SO_2 and NOx emissions and have the ability to combust different types of fuels. They also typically include a bag house or equivalent for particulate control and can meet future mercury emission standards. With pollution control equipment, CFB boilers are better able to operate at part-load and continue to meet their emissions performance. In any event, a CFB boiler will not want to operate at less than 60 percent of full output. At less than 60 percent of full output, a CFB will have trouble meeting the emission requirements of its air permit. At the present time, an investor can get comfortable with an investment in CFB boilers with a view that a CO_2 tax will be small and that the future price of power will recover to the replacement cost of a combined-cycle power plant. This combined-cycle revenue requirement calculation was developed in Chapter 2 of the book. Since combined-cycle power plants can get permitted and built, it is important to consider this calculation when evaluating any new or existing power plant investment.

Operating coal plants that already have SO_2 scrubbers and NOx control should be able to meet the proposed future 90 percent removal requirement for mercury control. If an operating coal plant doesn't have these control technologies, it may be forced to install activated carbon injection. A number of smaller coal plants don't have these controls, and as a result it would be uneconomic to retrofit them. Plants of this type might operate only on a "limited hours of operation" basis to provide grid support, as mentioned earlier.

Depending on local fuel conditions, they may also cofire biomass or, if small enough, switch solely to biomass. There is also a potential technology under development that would allow the cofiring of municipal solid waste with coal-fired power plants. Coal-fired plants that adopted this technology might also be able to qualify for RECs for the percentage of power that was generated with this waste or biomass fuel. The key issue will be the delivered cost of this fuel on a $/MMBtu basis and the value of any renewable energy credits that could be claimed. The emissions of the boiler will also be affected by this fuel. The goal will be to try to continue to meet the upper bound on the existing air permit. This strategy would result in only an administrative change and would avoid having to file a major change to the existing air permits and a public comment period. Use of biomass could help an existing coal-fired power plant reduce its SO_2 and CO_2 emissions but not its NOx emissions. Some type of particulate control will also be required. The future absolute emissions requirements of the EPA and local conditions, including attainment and nonattainment standards of key pollutants, will also determine the level of emission controls required.

Operating coal plants are facing a situation where the price of coal has taken off due to the need for both thermal and coking coal in India and China. Coal mines in both India and China can't keep up with their respective countries' internal demand for coal. Coal in India tends to be of low heat content, and as a result is expensive to ship for a long distance. Coal from the U.S. Central Appalachian region is actually being shipped to China for use in coking coal furnaces. A coal export terminal on the West Coast, which would export coal from the Powder River Basin, is also under consideration. Coal plants are also competing against the low price of natural gas. In the past, coal plants operated in a world of natural gas at prices above

$6/MMBtu and delivered coal of $2/MMBtu. This higher cost for natural gas provided coal plants with an added energy margin.

COAL TECHNOLOGIES

There are currently three main coal-fired power plant technologies, circulating fluidized bed (CFB), pulverized coal (PC), and integrated gasification combined cycle (IGCC). Due to the difficulties in obtaining new air permits, the low cost of natural gas, and the overall reduction in demand for power, siting new coal plants of any type is difficult. There will be a number of restructuring opportunities with operating coal plants with existing air permits.

CFB boilers have a niche in the combustion of low-quality fuels like waste coal or petroleum coke. Unlike PC boilers, the largest single-unit CFB tends to be 300 mW in size. Single-unit PC boilers can be as large as 700 mW in size and enjoy economies of scale. CFBs can enjoy diseconomies of scale since the opportunity fuels that they burn can be much lower in cost than the premium fuel that a PC boiler burns. These fuels might also have a tipping fee or a payment to the fuel purchaser. Depending on existing fuel-handling infrastructure, CFBs also have the ability to burn different fuels. Bubbling fluidized bed boilers, a CFB derivative, have been used recently to combust biomass. The all-in cost of a CFB boiler, including the engineering, procurement, and construction, would be in the range of $3,000/kW. This cost would make a new CFB uneconomic in the current power market, and it would be next to impossible to obtain an air permit due to CO_2 concerns. Foster Wheeler and Alstom are two manufacturers of CFB boilers.

Like a traditional PC boiler, a typical CFB project would include one or more steam turbines, generator, condenser, and cooling tower. Other components would include a fuel storage building, water treatment building, administration building, and a maintenance and warehouse building. In a CFB boiler, air and fuel are fed into the bottom of the combustion chamber and circulated continuously until combustion is complete. A fluidized bed is a suspension of coal and limestone in a flow of hot air. As it burns, limestone reacts with sulfur in the coal to form sulfate. The only by-product of the combustion process is a benign alkaline ash that can be used in mine reclamation and to reduce acid drainage from coal mines. Continuous circulation of solids provides longer particulate residence time, resulting in efficient fuel combustion and emissions control. In the CFB boiler, air, limestone, and fuel are fed into the combustion chamber and the solids are circulated continuously and combusted until they are small enough to be carried out with the flue gas. Continuous circulation of solids provides longer particulate residence time, resulting in efficient fuel combustion. Low combustor temperature and introduction of limestone into the combustor chamber reduce emissions.

Limestone is used to control CFB boiler SO_2 emissions. Limestone is typically delivered by truck, stored in a silo, and fed into the boiler. Aqueous ammonia is fed into the flue gas to control NOx emissions. This process is known as selective noncatalytic reduction (SNCR). Flue gases exiting the CFB pass through baghouses to remove particulates prior to discharge from the plant stack. The use of fabric filters in the baghouse reduces particulate matter to a very low level. Fly ash and bottom ash are collected and conveyed to an ash silo. Ash from the silo is discharged into trucks for disposal at offsite locations.

As a result of the limestone injection for ash control, the ash by-product produced by the CFB is high alkaline. It is very good for both active and abandoned coal mine reclamation. The ash can neutralize acid mine drainage from waste coal sites. Most newer CFB and PC plants also have a zero water discharge requirement. This means that no water from the plant would be discharged into local rivers or streams.

PC boilers are drum-type boilers equipped with air preheaters, soot blowers, and fans. Like a CFB, they are sized to provide an adequate amount of steam for the steam turbine. PC boilers are typically equipped with three mills capable of reducing coal to a required fineness and mesh size. PC boilers are typically designed with the ability to operate with one mill out of service. Like a CFB, PC boilers are usually equipped with a baghouse or electrostatic precipitator to collect and control fly ash. Modern PC boilers also have a scrubber for SO_2 control and SCR for NOx control. Foster Wheeler and Alstom are two manufacturers of PC boilers. Both of these firms have maximized their use of Chinese-manufactured equipment. A number of Chinese manufacturers are challenging both of these firms in both PC and CFB boilers. For a PC plant built in the United States, estimates from the Department of Energy's (DOE's) Energy Information Administration (EIA) are \$3,167/kW for a 650-mW single-unit facility. As with a CFB boiler, it would be extremely difficult to obtain an air permit for a facility of this type in the United States due to future CO_2 concerns.

IGCC allows for the use of coal in a gasification process. Coal is gasified to produce a synthetic gas, which is used to fire a gas turbine engine. As in a traditional GTCC, exhaust heat from the gas turbine engine is used to produce steam to drive a steam turbine. IGCC also has the advantage of potentially controlling CO_2 emissions by the use of a shift reactor. The CO_2 stream from the synthetic gas would be shifted out to create hydrogen. The hydrogen would be used to fire the gas turbine engines, and the CO_2 would be buried underground. Most gas turbine engine manufacturers have limited experience with burning 100 percent hydrogen on a large scale. CO_2 storage on a large scale has not yet been proven and will be expensive. Other IGCC projects have studied the use of selling their CO_2 to a CO_2 products pipeline. CO_2 has a number of uses, including for enhanced oil recovery. The challenge is that there is not a large enough market to use if all of the CO_2 was captured from every operating power plant. This is the reason that underground sequestration of CO_2 has to be studied. The enhanced oil recovery market use of CO_2 would be quickly overwhelmed if a cost-efficient method for capturing CO_2 were found.

The challenge that IGCC faces is its cost. Unlike wind and solar projects, IGCC plants don't qualify for the investment tax or production tax credit. The U.S. government has been offering outright grants to IGCC projects in an effort to try to make them competitive. States are not required to buy power from IGCC facilities under their REC program. Like other coal projects, the low cost for natural gas makes it hard for IGCC facilities to compete. The DOE EIA shows the cost for a 600 mW IGCC facility at $3,565/kW, which is below the projected cost for projects under development. When CO_2 control is considered, the output drops to 520 mW and the cost increases to $5,348/kW.

The state of Illinois encouraged the development of an IGCC plant by forcing the local utilities to buy power from a project that uses Illinois coal and fully sequestered or sold its CO_2 emissions. The proposed IGCC Taylorville Energy Center (TEC) in Illinois is estimated to have an all-in cost of $3.52 billion for a 602-mW power plant. This works out to $5,847/kW, above the DOE EIA IGCC estimate. TEC attempted to contract with a CO_2 pipeline that would move the plant's CO_2 to the Gulf Coast. This concept didn't work since the pipeline was not able to obtain all of its right-of-ways. As of January 14, 2011, the Illinois state senate had voted against approving the plant since it would significantly raise the cost of power. At the time of this writing, the TEC is stalled.

Duke Energy's Edward's Port IGCC facility is the only IGCC facility that is currently under construction. Duke originally estimated that the project would cost $1.985 billion. The most recent estimate, as the project is now 57 percent complete, is $2.88 billion. Numerous other IGCC projects have been deferred or canceled. Xcel Energy announced in 2007 that it was indefinitely deferring its plans to build an IGCC facility in Colorado.

SUMMARY

The technical and economic issues of coal plants have to be understood in order to evaluate renewable power plants. Like renewable power projects, coal power projects face a challenge from abundant supplies of shale gas. Some investors believe that it will be difficult for coal power plants smaller than 500 mW in size to continue to compete. Obtaining an air permit for a new coal power plant is next to impossible, making some operating coal plants interesting investment opportunities.

Chapter 13 reviews biomass, which can be cleanly burned in specially designed boilers or cofired with coal.

Biomass Energy and Biomass Power Plants

"The greatest Oaks have been little Acorns."
—Thomas Fuller's *Gnomologia,* 1732

Biomass is defined as a renewable organic material that can be used to produce energy. Most U.S. states allow biomass-based power plants to qualify for renewable energy credits (RECs). Unlike a wind or solar power plant, a biomass power plant can be a seven-day-a-week, 24-hour-a-day resource. On an annual basis, a biomass power plant can have an availability and capacity factor over 90 percent. Like other renewable power sources, biomass power plants also have to compete against low-priced natural gas.

Biomass energy is derived from various types of vegetation, trees, branches, roots, bark, and animal or human waste. In China and India, various nonedible plants, like bagasse, are collected and burned in power plants to create energy. Bagasse is a by-product from sugar cane production that is used as a boiler fuel. Other types of biomass that are not used for food can be collected for mulch, fertilizer, or burned in power plants as fuel.

Today, biomass, which is safe, nonhazardous, and nonlethal fuel, is burned in power plants at very high temperatures and can produce a great deal of heat or energy. The burning of biomass fuel does create carbon dioxide (CO_2), nitrous oxide (NOx), and particulate emissions.

WOOD WASTE

Today, wood waste is one of the most widely used forms of renewable fuel in the United States for homes, schools, factories, prisons, and power plants in general. In 2010, biomass from wood waste exceeded the total energy produced from all our nation's hydropower plants. Biomass fuels produce more than 7,700 megawatts (mW) of electricity. This is because many thousands of small U.S. power plants, plus large U.S. power plants, can use wood waste according to the U.S. Department of Energy's Federal Energy Management Program.

Less than 50 percent of each tree ends up in finished lumber, furniture, flooring, cabinetry, doors, walls, stairs, toys, children's climbing frames, beds, frame buildings, and construction sites. Therefore, all the rest of each tree results in underutilized products or is dumped in landfill sites. Much wood waste is derived from construction sites for new buildings where many pieces of wood are used to frame the building, or as long wooden molds for cement to be poured in to harden. The greater the number of lumber mills and furniture manufacturers or the more new construction sites in a city, the more wood waste can be expected to be available for wood waste to energy power plants. The challenge for a biomass power plant is to locate as close as possible to these supplies in order to reduce transportation cost

On a $/MMBtu basis, wood waste is often more expensive than coal. Wood waste has a typical as-received heat content of only 4,500 Btu/lb. Assuming a delivered price of $40/ton, this converts to:

$$\$40/\text{ton} \times 1\,\text{ton}/2{,}000\,\text{lb} \times \text{lb}/4{,}500\,\text{Btu} \times 1{,}000{,}000\,\text{Btu}/\text{MMBtu}$$
$$= \$4.44/\text{MMBtu}$$

This delivered price is well above the delivered price of coal. It is the range of the delivered price of natural gas. Unlike a natural gas plant with a 7,000 Btu/kWh heat rate, a typical biomass plant will have a heat rate of 13,500 Btu/kWh. This converts to an energy price of:

$$13{,}500\,\text{Btu}/\text{kWh} \times \text{MMBtu}/1{,}000{,}000\,\text{Btu} \times 1{,}000\,\text{kWh}/\text{mWh}$$
$$\times \$4.44/\text{MMBtu} = \$59.94/\text{mWh}$$

This compares to the energy price of a new combined-cycle plant of only:

$$7{,}000\,\text{Btu}/\text{kWh} \times \text{MMBtu}/1{,}000{,}000\,\text{Btu} \times 1{,}000\,\text{kWh}/\text{mWh}$$
$$\times \$5/\text{MMBtu} = \$35/\text{mWh}$$

According to the DOE EIA, the capital cost for a biomass project would be in the range of $3,860/kW, which is substantially above the capital cost for a new gas-fired combined-cycle power plant.

Depending on the quality of the wood waste, the cost of disposal can be avoided and wood waste may have a negative price. With the increased interest in biomass-fired power plants, this has become a rare situation. Wood waste collected on federal lands after lumbering may reduce the risk of forest fires.

Wood waste can be used for space heating, process heat, or direct electricity production. The most common industrial use of wood for energy production is when steam is produced in a boiler using standard stoker technology or bubbling or fluidized bed boiler technology. Wood waste can also be blended with coal in order to decrease the level of emissions of NOx and sulfur dioxide (SO_2) that the burning of ordinary coal produces. It is possible to retrofit most existing coal plants for cofiring with wood, and this can significantly decrease the toxic emissions of SO_2 and NOx.

ECONOMICS OF BIOMASS

At the present time, it is difficult to make the economics work for gasification using biomass. Biomass projects tend to be smaller in size and don't benefit from economies of scale.

There are many U.S. federal agencies that can use wood as a way to use energy-saving performance contracts to finance their energy projects that enable U.S. government facilities to reduce their energy use and costs without requesting congressional appropriations to fund these projects. All over the world, U.S. federal facilities can use technology-specific biomass and alternative methane fuel (BAMF), which offers private-sector expertise specifically geared to using renewable BAMF resources.

In Vermont, 25 schools over the past 15 years have been using 8,000 tons of wood chips per year to heat their schools. Similarly, in Maryland, the Department of Corrections facilities have cut their fuel costs by 63 percent by producing their own power by using wood chips for heating all their prisons instead of coal or gas. In the case of both the Maryland prisons and the Vermont schools, the power plants involve burning wood chips to boil water, which produces steam to power two condensing steam turbines rated at 1.9 mW. The prison was expanded to 3,100 beds, resulting in a 60 percent increase in energy demand. The original system continues to service the expanded facility with reliable low-cost energy.

In another case, the Central Michigan University (CMU) campus at Mount Pleasant was given a wood-fired energy system as a retrofit to an existing natural gas–fired system at a cost of $3.6 million. This resulted in a payback period of less than four years. The conversion included the addition of a boiler rated for 50,000 lb per hour of steam. A 1-mW steam turbine generator provides electrical power and serves as a pressure-reducing valve for steam that is used downstream for heat, air conditioning and hot water. The CMU system is designed to burn 43,700 tons of wood tree chips per year. The chips are harvested from low-grade wood supplies within a 50-mile radius of the campus. CMU estimates that the school is saving $1 million per year through the reduced cost of fuel and is injecting approximately $1 million per year into the local and state economy from wood harvesting and processing operations.

SUMMARY

Biomass is a renewable technology that enjoys a high capacity factor. It is difficult to find enough wood waste to supply a power plant larger than 50 mW in size. Ohio Edison attempted to repower an approximately 300-mW coal plant and was unsuccessful in locating enough biomass that could be economically delivered to the plant. Transporting wood waste more than 50 miles is not economical due to its low as-received heat content of approximately 4,500 Btu/lb. Depending on a power plant location and its cost of coal, existing coal-fired power plants can often be cofired with 10 percent wood residues with only minor plant modifications.

Certain states are making it difficult for biomass projects to collect RECs if they don't have a baghouse or fabric filter for particulate control. This can occur even if the biomass facility is in compliance with local air regulations. REC payments to biomass projects are critical due to their capital and variable cost disadvantages to natural gas–fired power plants.

In Chapter 14, we discuss nuclear power energy plants.

CHAPTER **14**

Nuclear Power Energy Plants

*To the making of these fateful decisions, the United States pledges
before you—and therefore before the world—its determination to
help solve the fearful atomic dilemma—to devote its entire heart
and mind to find the way by which the miraculous inventiveness of
man shall not be dedicated to his death, but consecrated to his life*
—President Eisenhower's speech to the United Nations on his
"Atoms for Peace Program"

On March 11, 2011, Japan experienced a major earthquake that was rated as 9 on the Richter scale, the highest level ever recorded in Japan. In fact, it was one of the most powerful earthquakes recorded worldwide during the past 100 years. This earthquake caused a huge tsunami in the ocean right off the Japanese shore, and together they caused a nuclear meltdown in three nuclear power plants at Fukushima Daiichi nuclear station and also negatively impacted two other nuclear power plants. The result of this combination of natural disasters was the successive failure of the three safety mechanisms that Japanese nuclear plants relied upon. These power plants, built in the 1970s, had been built to withstand earthquakes because Japan is located on top of a well-known earthquake zone. However, none of the earthquake tests on Japan's nuclear power plants had been rated as high as 9. Also, no nuclear power plants had been built with testing of the full impact of a huge nearby simultaneous tsunami.

Global TV videos of massive destruction were being broadcast continually as international nuclear rescue teams were rushed to the site and the local population was evacuated. Although the Japanese government's municipal and national measures to calm the public and to prevent panic in this disaster were extraordinarily successful (based on many years of Japanese public earthquake training), the deaths numbered over 25,000.

Yet unrecovered bodies of people drowned in the ocean and buried under huge catastrophic levels of building and infrastructure debris led to a wide range of estimates of fatalities, injuries, and damage. The radioactivity of the air, water, dust, and iodine in the seawater were thousands of times higher than the "safe level."

GLOBAL IMPACT OF JAPAN'S THREE NUCLEAR PLANT MELTDOWNS

On January 6, 2011, China announced publicly that it would begin a major process of reusing virtually its total of spent uranium fuel. However, following the Japanese nuclear crisis of March 11, 2011, China—which had 79 new nuclear power plants to be constructed—announced that it had temporarily halted until the results of its scientific investigations of nuclear power plant safety (under the new earthquake, tsunami, and nuclear plant meltdowns in Japan and elsewhere) could be more definitively evaluated.

In the European Community (EC), the Minister of the Environment declared that there would be a three-month investigation of all 143 nuclear power plants in the EC to determine whether existing nuclear power plants were vulnerable to earthquakes, tsunamis, airplane crashes, terrorist or cyber attacks, or major water problems or coolant problems. In Germany, Chancellor Angela Merkel had recently proposed an automatic extension of 12 years added to the life of the 17 German nuclear reactors; she was suddenly forced, three days after the March 11 nuclear crisis, to declare a three-month contractors' moratorium on the extension of any nuclear power plants until an in-depth scientific study could be conducted and completed. Within days, two older nuclear German nuclear power plants were discontinued, and other European nations were testing their own nuclear power plants for safety. By May 31, 2011, Merkel declared that Germany would close all its nuclear plants.

On May 26, Switzerland, where 40 percent of the entire nation's electricity is derived from nuclear plants, declared that the Swiss cabinet recommended to the Swiss Parliament the end of nuclear power as an energy source.[1] Italy and other nations around the world also put their proposed nuclear power plant construction projects on hold, and many instantly set in motion vast inspections of nuclear plant safety. Global financial markets immediately lowered the stock price of nuclear power corporations, and the

---

[1] Goran Mijuk and Markus German, "Swiss Move to End Nuclear Era," *Wall Street Journal*, May 26, 2011.

rating agencies downgraded the credit rating of such companies around the globe.

Continuing global research studies of the nuclear damage in Japan three months after the event indicated ongoing damage from the original events. They now suggested that they had not successfully contained the nuclear danger to the total of five nuclear plants, and new ongoing dangers were being discovered in the water, hydrogen, and remaining nuclear fuel still in the plant.

The already high cost estimates for constructing new nuclear power plants versus other forms of energy power plants suddenly rose to take into account the many anticipated extra costs of earthquake and tsunami protection that would be required worldwide, and also the new government safety precautions and wide range of new tests that would be required for final authorization of new nuclear power plants. There was also a spike in projected costs because of the widely anticipated cancellation of many nuclear power plants in various nations.

France was a nation nearly 80 percent dependent on nuclear power plants for all its electricity needs, the United States was 19 percent dependent for all its electricity (and produced the most nuclear power–derived electricity of any nation), and Japan was 20 percent reliant on its nuclear power plants for its total electricity needs. These were three nations most directly affected by this international nuclear crisis.

However, well over 30 nations were dependent on nuclear power for significant portions of their electricity, as well as for energy for hundreds of naval ships, icebreaker ships, and aeronautic space rockets and many scientific experiments in medicine, new materials, and mining and numerous other industries, were also thrown into financial and energy safety uncertainty.

Even before the Japanese nuclear plants disaster, nuclear power plants were already one of the most (if not the most) expensive sources of electricity on a cost per plant/kilowatt basis, compared to not only various fossil fuels such as coal, oil, gas, natural gas, shale gas, and liquefied natural gas, but also the various renewable energy sources, such as hydropower, solar power, wind power, thermal power, biomass, wave power, and tidal power. The cost of mining the uranium or other radioactive ore such as plutonium and thorium (a less radioactive ore), as well as the cost of then transforming ore into a safe form of "yellow cake" or other transportable substance, were really only the first basic costs, compared to the total all-in cost of obtaining all government regulatory permissions, designing the plant, and constructing the plant, and final completion tests and numerous safety evaluation procedures, which were often subject to very long delays and changes.

The costs of fuel once the plant was fully built and approved were relatively small, since the nuclear fuel lasted so long and could generate electricity for decades. Yet, all the initial costs of the nuclear power plant were huge and the scale of nuclear plants tended to be the highest because it was baseline power available 24/7 for three to six decades. The result was that the cost comparisons of other types of power plants, which were much smaller than nuclear power plants, were much cheaper to build but often required very high continual ongoing fuel expense. The fact that the cost comparisons for one electric kilowatt were often not truly "apples to apples" equivalencies for nuclear power, because of the huge up-front costs that nuclear power plants required, enabled large-volume *discounts per kilowatt.*

COMPARATIVE COSTS OF ENERGY

To do a comparative evaluation of a typical "Updated Estimate of Power Plant Capital and Operating Costs" calculated by the U.S. Energy Information Administration (EIA), we find that "conventional natural gas combined cycle" is stated as of 2010 to have a nominal capacity of 540,000 megawatts (mW), has an "overnight capital cost" of $978/kW) with a "fixed operating and maintenance (O&M) cost (2010$/kW) of $14.39. By contrast, a "uranium dual-unit nuclear plant" has a "nominal capacity" of a large 2,236,000 mW, but a low variable O&M cost of $2.04/kW. See Table 14.1.

The basic updated capital cost estimates for electricity-generating plants have been repeatedly evaluated by both the EIA and various private research organizations and corporations like the National Economic Research Association (NERA), at different times every year.

KEY TO THE EIA COST ESTIMATES

The following is a key to the assumptions and sources used in calculating a range of "levelized costs for generating electricity from different fossil, nuclear, and renewable energy technologies brought online in 2015." Levelized cost of electricity is a measure often used by analysts to compare and evaluate the relative costs and competitiveness of different electric power generating technologies.

Different major research organizations use somewhat different key assumptions and sources in calculating a range of "levelized costs for generating electricity" from different fossil, nuclear, and renewable energy technologies. See Tables 14.2 and 14.3.

TABLE 14.1 Updated Estimates of Power Plant Capital and Operating Costs

	Plant Characteristics		Plant Costs		
	Nominal Capacity (kilowatts)	Heat Rate (Btu/kWh)	Overnight Capital Cost (2010/kWh)	Fixed O&M Cost (2010$/kW)	Variable O&M Cost (2010/MHz)
Coal					
Single-unit advanced	650,000	8,800	$ 3,167	$ 35.97	$ 4.25
Dual-unit advanced PC	1,300,000	8,800	$ 2,844	$ 29.67	$ 4.25
Single-unit advanced PC with CCS	650,000	12,000	$ 5,099	$ 76.62	$ 9.05
Dual-unit advanced PC with CCS	1,300,000	12,000	$ 4,579	$ 63.21	$ 9.05
Single-unit IGCC	600,000	8,700	$ 3,565	$ 59.23	$ 6.87
Double-unit IGCC	1,200,000	8,700	$ 3,221	$ 48.90	$ 6.87
Single-unit IGCC with CCS	520	10,700	$ 5,348	$ 69.30	$ 8.04
Natural Gas					
Conventional NGCC	540,000	7,050	$ 978	$ 14.39	$ 3.43
Advanced NGCC	400,000	6,430	$ 1,003	$ 14.62	$ 3.11
Advanced NGCC with CCS	340,000	7,525	$ 2,060	$ 30.25	$ 6.45
Conventional CT	85,000	10,850	$ 974	$ 6.98	$14.70
Advanced CT	210,000	9,750	$ 665	$ 6.70	$ 9.87
Fuel cells	10,000	9,500	$ 6,835	$350.00	$—
Uranium					
Dual-unit nuclear	2,236,000	N/A	$ 5,335	$ 88.75	$ 2.04
Biomass					
Biomass CC	20,000	12,350	$ 7,894	$338.79	$16.64
Biomass BFB	50,000	13,500	$ 3,860	$100.50	$ 5.00

(continued)

TABLE 14.1 (Continued)

	Plant Characteristics		Plant Costs		
	Nominal Capacity (kilowatts)	Heat Rate (Btu/kWh)	Overnight Capital Cost (2010/kWh)	Fixed O&M Cost (2010$/kW)	Variable O&M Cost (2010/MHz)
Wind					
Onshore wind	100,000	N/A	$ 2,438	$ 28.07	$—
Offshore wind	400,000	N/A	$ 5,975	$ 53.33	$—
Solar					
Solar thermal	100,000	N/A	$ 4,692	$ 64.00	$—
Small photovoltaic	7,000	N/A	$ 6,050	$ 26.04	$—
Large photovoltaic	150,000	N/A	$ 4,755	$ 16.70	$—
Geothermal					
Geothermal—dual flash	50,000	N/A	$ 5,578	$ 84.27	$ 9.64
Geothermal—binary	50,000	N/A	$ 4,141	$ 84.27	$ 9.64
MSW					
MSW	50,000	18,000	$ 8,232	$ 373.76	$ 8.33
Hydro					
Hydroelectric	500,000	N/A	$ 3,076.00	$ 13.44	$—
Pumped storage	250,000	N/A	$ 5,595.00	$ 13.03	$—

TABLE 14.2 Levelized Cost of Electricity from Fossil and Nuclear Technologies (2010$)

Plant Nameplate Capacity	Units mW	Supercritical Pulverized Coal 600	Coal IGCC 600	Coal IGCC-CCS 480–520	Natural Gas CC 400	Nuclear 1100–1350
Overnight capital cost	$/kW	2,800–3,400	3,200–3,800	5,000–6,500	1,000–1,300	5,000–6,000
Fixed charge rate (a)	%	21.3%	23.1%	18.0%	15.9%	18.9%
Fixed O&M cost	$/kW-yr	35.97	59.23	69.3	14.62	88.75
Variable O&M cost	$/MHz	4.25	6.87	8.04	3.11	2.04
Heat rate	Btu/kWh	8,800	8,700	10,700	6,430	10,500
Capacity factor	%	85%	80%	80%	50–87%	80–90%
Fuel price (levelized)	$/MMBtu	1.60–2.70	1.60–2.70	1.60–2.70	4.00–6.75	0.8
Fuel price (levelized)	$/mWh	14.1–23.9	13.9–23.6	17.1–29.1	26–43.5	8.2
Construction period	Years	4–5	4–5	4–6	3	6–7
Levelized cost of electricity	$/mWh	103–130	135–164.5	163.4–214	52–97.5	141–184
Incentives (levelized) (b)	$/mWh	n/a	21.1–25.1	25.7–33.4	n/a	49.9–17.5
CO_2 Cost (levelized)	$/mWh	16.6–49.8	16.2–48.5	3–5	6.8–20.5	n/a
Levelized cost of electricity with incentives and CO_2	$/mWh	120–180	130–188	141–190	59–118	91–167

TABLE 14.3 Levelized Cost of Electricity from Renewable Energy Technologies (2010$)

Plant Nameplate Capacity	Units	Wind (Onshore) 50–100	Geothermal 50	Biomass CFB 50	Large Solar PV 1–100	Solar Thermal (b) 50–100
Overnight capital cost	$/kW	2,000–2,500	3,000–10,000	3,800–4,300	3,000–4,500	4,700–6,800
Fixed charge rate (a)	%	9.80%	11.00%	11.40%	9.80%	10.50%
Fixed O&M cost	$/kW-yr	28.07	84.27	100.5	16.7	64
Variable O&M cost	$/MHz	n/a	9.64	5	n/a	n/a
Heat rate	Btu/kWh	n/a	n/a	13500	n/a	n/a
Capacity factor	%	25–45%	85%	80%	20–28%	27–43%
Fuel price (levelized)	$/MBtu	n/a	n/a	1.88–4.06	n/a	n/a
Fuel price (levelized)	$/MWh	n/a	n/a	25.3–54.8	n/a	n/a
Construction period	Years	1	3–4	3–4	1	2–3
Levelized cost of electricity	$/mWh	57–125	65–169	107–144	126–260	147–328
Incentives (levelized) (c)	$/mWh	21	21	21	35.8–75.3	39–90
CO_2 cost (levelized)	$/mWh	n/a	n/a	n/a	n/a	n/a
Levelized cost of electricity	$/mWh	36–104	44–148	86–123	90–185	108–238

NUCLEAR POWER PLANTS' 50 YEARS OF ELECTRICITY GLOBALLY

Globally, nuclear power plant history has experienced dramatic shifts over the past 50 years due to very strong political, military, scientific, financial, and social confrontations. Great Britain was the original pioneer of the peaceful use of nuclear power reactors for public energy. However, different nations created a range of different types of nuclear power reactors. These different nations also had different conceptions of what the required safety standards of nuclear power plants should be and all the safety protection procedures to prevent a meltdown and different containment procedures after a nuclear meltdown occurred.

Over the years, a few nuclear accidents and meltdowns occurred: the Chernobyl plant in the Soviet Union, the Three Mile Island plant in the United States, plus accidents at nuclear submarines and ships, and so on. All of these accidents galvanized Greenpeace and other international antinuclear movements to launch major national political marches, demonstrations, and sometimes violent attacks against governments for two decades.

In 2003, in response to nuclear protests, the British Labour Government passed an act to stop all nuclear power plant construction. However, by 2006, the British Labor Government plus the atomic scientific community and the financial markets decided to make the nuclear industry a center of a new national green initiative to reduce Britain's carbon footprint and very significantly to remove carbon dioxide from the British lungs, climate, air, water, waste treatment, roads, transport, buildings, and the total environment.

France continued to derive nearly 80 percent of its entire electric energy from nuclear power plants. The United States continued to derive 19 percent of its electric energy from its nuclear power plants, and many other nations continued to rely heavily on their nuclear power plants or to buy nuclear power from other nations. Germany had relied on nuclear power plants for 47 percent of its electricity and, in fact, the prime minister, Angela Merkel, had continued extending the lives of Germany's nuclear power plants for decades, even days before Japan's nuclear meltdown crisis.

Almost immediately following the Daiichi Japan level 9 earthquake, the tsunami, and triple nuclear meltdown disaster, Merkel announced a halt to all new nuclear work until there was an in-depth scientific and engineering analysis of all Germany's nuclear plants. The United States and many other nations also ordered the most stringent tests on all their nuclear power plants to be certain that they could all withstand a 9-rated earthquake on the Richter scale, a volcanic eruption, a jet aircraft crash, a giant meteor, acts of terrorism, or natural disasters like tornadoes, cyclones, or hurricanes.

Even when the simultaneous giant Japanese earthquake, tsunami, and three nuclear power plants meltdown occurred in March 2011, the British Coalition Government of Conservatives plus Liberal Democrats publicly declared they would continue the British nuclear power industry. While they would take extra safety precautions and conduct careful scientific studies on their own, and in conjunction with other nations, the U.K. State said it would "go forward" with their master plan for a new generation of nuclear plants, fast breeder reactors, and new nuclear equipment for reprocessing of used nuclear material.

While Germany and Switzerland decided to abandon their nuclear power option, other, faster-growing nations in Asia and Europe proclaimed their conviction that they would increase efforts to develop advanced nuclear "third generation" and "fourth generation" nuclear reactors as a vital part of their total energy mix.

REQUIRED UP-FRONT PAYMENT FOR NUCLEAR WASTE DISPOSAL BEFORE A NEW PLANT'S APPROVAL

One major recent innovation of new nuclear power plant planning in a number of nations is the requirement of a "geological disposal facility" (GDF), which is a guaranteed financial commitment (up front) to pay for the safe disposal of the nuclear waste created by each nuclear power plant. This prepayment program is especially reasonable because the cost of new advanced methods of creating nuclear power will have to be originally factored into the cost of their "reprocessing" their "spent" or "waste nuclear material" one, two, or many times together with their potential range of ultimate nuclear waste disposal costs.

As a result of financial planning and budgeting in advance, the extra added life span of the original nuclear fuel that will be reprocessed, and the safe number of times it can be reprocessed, while it will cost more initially than it would have originally, the nuclear power plant will have a prepaid right to reprocess this nuclear fuel. This lowers the average all-in disposal cost of spent nuclear fuel by reducing the ultimate amount of totally spent nuclear fuel. In other words, as the French, the Japanese, and now the Chinese have discovered, the more times that nuclear fuel can be safely reprocessed and used to create new energy, heat, light, or power, the ever higher total amount of nuclear fuel that nation has ultimately created and used. Conversely, the ever-lower final amount of used fuel that will be left over, in the end, will cost significantly less to be safely disposed of. The reduction of that nuclear waste disposal cost for decades has been a huge and "very costly unknown in the nuclear power industry" as demonstrated by the two decades wasted nationally, regionally, and locally by numerous government regulatory agencies and legislators in the ultimately fruitless effort in the United States trying to site and test and then failing to get approved the "Yucca Mountain Nevada national nuclear disposal site" thousands of feet deep under solid rock, which was ultimately rejected by state, local, and federal law.

It is only by calculating accurately these total reprocessing costs of nuclear fuel and the total financial benefits in total reprocessed nuclear energy produced by each extra given ton of material that it can be accurately quantified. It is this new way of comprehensively evaluating the total costs and total benefits of the new reprocessing technologies for nuclear plants that can now bring down its lifetime cost of energy to become somewhat more comparable to coal, oil, gas, or other energies over the long term.

Nuclear power generates few greenhouse gas emissions into the earth's atmosphere, whereas oil and coal generate large emissions. Even traditional gas, and "shale gas" acquired through the fracking of shale rocks to extract natural gas, does produce CO_2 emissions at about half the rate of coal. This is less CO_2, but for nations that stated their intent to reduce greenhouse gas emissions to their absolute minimum, a switch to shale gas pollution emissions is still a carcinogen.

ASIA WILL LEAD THE NEXT SHIFT TO NUCLEAR POWER PLANT DEVELOPMENT

China, India, South Korea, and Russia are each significantly expanding their current total of nuclear power plants. China is building new nuclear energy plants by its own CPR 1000 design construction, plus joint ventures with Westinghouse/ToshibaAP, Russia's VVER nuclear plant design, South Korea's APR1400 nuclear plant design, and several other foreign national nuclear energy construction companies. China also is purchasing other nations' new nuclear power plants.

In 2010, India had 20 operating nuclear power plants. By 2011, it had six more new nuclear power plants under construction. India also had stated plans to build and/or purchase 40 more new nuclear power plants from foreign nations to be in operation by 2032. India has signed binational contracts for nuclear plants with nine foreign nations. The India–South Korea Civil Nuclear Cooperation Pact was first signed in 2009, and finally reaffirmed by both nations, India's president, Pratibha Patil, and South Korea's counterpart, Lee Myung-bak, on July 25, 2011, in Seoul, Korea.[2]

---

[2] http://newsdawn.blogspot.com/2011/7/india-south korea-civil-nuclear-pact.

South Korea's sales of its APR1400 nuclear power plants to the United Arab Emirates in 2009–2011 are forecast to lead to more sales of global nuclear power plants.[3] Edward Kee, vice president at NERA Economic Consulting in Washington, D.C., has spoken on the impact China and South Korea will have on the global nuclear industry.[4]

In his speech in Hong Kong in December 2010, Kee stressed that the most important issue for reactor designs is to

> . . . *get a lot of units built and into operation as fast as possible. This gets the design down the learning curve to lower costs and shortens schedules, but also stimulates additional sales from buyers who look for low risk and demonstrated success. While design features are important, market success is much more important.*[5]

GE and other nuclear corporations let themselves get tied up in highly complex U.S. Nuclear Regulatory Commission (NRC) licensing processes and legal and financial procedures, with the result that few major U.S. projects have started, while China, India, South Korea, and Russia are all building many plants.

The vital fact is that all these Asian nuclear power plant builders have their total governments' financial backing, and all state electric utilities have large numbers of continuing orders for new plants. None of the U.S. or European nuclear companies have anything like that. Thus, Asian nuclear firms gain continual experience and expertise and relationships with each other and many governments worldwide.

Kee makes a key point when he compares Asia to the United States and Europe:

> *Western vendors must cobble together a series of subcontractors and related agreements from unrelated commercial entities; each of these agreements adds cost to meet risk premiums and profit margins of subcontractors risk (as responsibility is shared between multiple commercial entities), and complexity (project management is more difficult due to multiple entities with multiple interests and contractual rights. . . . "*[6]

---

[3] Edward Kee, NERA Economic Consulting, Nuclear Power Conference Proceedings, Hong Kong, December 7–8, 2010.
[4] Kee, Nuclear Power Conference Proceedings.
[5] Ibid.
[6] Ibid.

CHINA'S NEW NUCLEAR REPROCESSING IS A VAST EXPANSION OF ATOMIC FUEL

Until recently, the total amount of nuclear fuel "reprocessing" done in India, China, the United States, and other nations has been limited, so that the maximum cost savings from one reprocessing has been estimated to have reduced the overall nuclear fuel cost by 15 to 25 percent.

However, in China's extraordinary announcement on January 3, 2011, Wang Junfeng, the project director of the Chinese National Nuclear Commission (CNNC), stated on China Central Television (CCTV) that at a remote site in the Gobi Desert in the Gansu Province, they had successfully begun to reprocess spent nuclear fuel material from light water reactors. Wang Junfeng then stated that China now had enough fuel to last 70 years, and the new technology could yield enough nuclear fuel to last for 3,000 years.

This incredible leap in nuclear fuel production stunned the world's nuclear community of scientists and engineers. Virtually all agreed that China could only conceivably achieve that unbelievable total expansion of nuclear fuel if it were to continually reprocess all its nuclear fuel many times over. But this alternative, many scientists fear, might cause potential health hazards that have never been faced before.

Potential Chinese global nuclear power plant health hazards resulting from excessive reprocessing of used radioactive nuclear material instantly reminded some experts of the Russian Chernobyl nuclear plant catastrophe. That most famous historic event caused by far the worst nuclear power plant disaster ever in 1986 at Chernobyl. That was because it spread nuclear clouds and radioactive ash destruction over thousands of miles of European, Asian, and Middle Eastern territory. It virtually stopped all new Russian construction of nuclear power plants for a decade. Its antiquated, unregulated, unsupervised, and extraordinarily dangerous plant became synonymous with "nuclear death." As a result, various older Russian plant reactors and their designs will never be accepted anywhere.

Yet, now Russia appears to have in part adapted some aspects of U.S., French, German, Indian, Canadian, and Scandinavian nuclear power plant designs. However, much more importantly, Russia has focused intently on rapid development of third- and fourth-generation nuclear plant designs specifically. These are fast breeder reactors that are usually "closed." Therefore, supposedly no human error occurs in running these power plants.

Most other advanced nations have also targeted their own scientific and technological development of these same third- and fourth-generation fast breeder reactors. Nevertheless, Russia is now back into significant nuclear plant construction and is selling its plants and joint-venturing with China and other governments.

It is specifically China's willingness to very aggressively joint-venture with Russia in new nuclear power plant design and construction that has concerned many of the best-known nuclear power plant contractors elsewhere because China is now the very largest customer, by far, in the entire world. All other nuclear power plant producers, mining companies specializing in uranium, thorium, plutonium, and other radioactive minerals anywhere in the world, are competing to sell their ore, their power plants, their technology systems, their nuclear fuel, or their spent nuclear fuel, or to enter into short-term or long-term contracts or joint-venture partnerships with China. The reason is that China has proclaimed its firm plan and clear intention to build 90 to 110 nuclear power plants, including those for its own use and for sale to other buyers or other nations along with very long-term service contracts.

In this way, China could now attempt to be positioned to be able to set world standards for nuclear power plants. China could drive their basic costs of construction down to rock bottom and also could be able to gain the greatest number and fastest-achieved "experience curve" benefits of any nuclear power plant producer in the world.

Because China did not embroil itself in major wars over the past two decades, nor did it blow up its own financial securities markets, it is in a better financial position than is the United States to partner with other nations in their nuclear power plant plans and many actual construction projects.

Westinghouse Corporation in the United States has formed a highly successful joint venture with Toshiba. Together, they have created what is currently considered the most advanced nuclear power plant in the world. This is the best third-generation fast breeder reactor. It is also said by some experts to possess certain elements that are capable of being used in building a fourth-generation nuclear power reactor. These Westinghouse/Toshiba power plants have been sold to China and to various other nations. This Westinghouse/Toshiba technological platform is the main reason that the United States is still able to compete in the world markets.

However, since the United States has not been yet been given final regulatory approval to be able to complete a brand new nuclear plant in the United States for 30 years, it has only been able to continue to work on trying to develop one new nuclear power plant and also upgrade its existing 100-plus nuclear plants. The United States also has joint-ventured with major nuclear power plant producers in other nations and assist with financing for various plants abroad.

For the past decade, China has been continually making million-pound purchases of uranium on the spot market for its planned giant 80-gW nuclear reactor. As a direct result, China drove the spot price up

$62.50/pound from last year. Reporters at *Fuel Cycle Week* believe that China's announcement is not so much a breakthrough as the start of development of a commercial-sized reprocessing plant of 800- to 1,000-tonne/year that uses French technology."[7]

Last July, China achieved criticality at its first prototype fast neutron reactor. The Chinese experimental fast reactor is expected to reach a thermal capacity of 60 mW and produce 20 mW of electric power. Developed by the China Institute of Atomic Energy, it is the first sodium-cooled fast reactor in the country. The reactor was built in collaboration with four different Russian nuclear development centers: the Kurchatov Institute, NIKIET, OKB Gidropres, and OKBM Afrikantov.

According to *Fuel Cycle Week*:

> *It appears that China has set aside plans for 600 MW Chinese design in favor of buying two BM800 fast reactors from Russia for Sanming-1 and 2. The project is expected to break ground in August 2011. A bilateral program on fuel cycle for fast reactors is part of the effort. China will build a nuclear city around the two reactors to house construction workers, reactor operators, and support services.*[8]

SUMMARY: NUCLEAR POWER FACES A CAPITAL COST AND ONGOING LOCAL APPROVAL CHALLENGE

The future of global nuclear power plants today is a far more complex forecast than it would have been before the 2011 Japanese earthquake, tsunami, and three nuclear power plant meltdowns. The entire nuclear power industry is under intense nationally or internationally mandated investigations to test whether any of their nuclear power plants are at risk from a level 9 earthquake.

Certain facts concerning the future of nuclear power plants and the future of other nuclear projects do raise important questions for investors in various nations.

First, there are nuclear plants already built on barges, ready to be installed in the Arctic Sea on the north coast of Russia that will power small new urban developments. These are forecast to be part of new Russian,

---

[7] *Fuel Cycle Week*, No. 358, December 6, 2010, http://fuelcycle.blogspot.com/.
[8] Dan Yurman, "China Now Reprocessing: A Beginning, Not a Breakthrough," *Fuel Cycle Week* 10, no. 406 (January 6, 2011), http://fuelcycle.blogspot.com/.

Norwegian, U.S., Canadian, and multinational oil, gas, and mineral exploration and commercial exploration of huge sectors of the Arctic Ocean as a result of new conclusions of 30 years of territorial, economic, and financial negotiations worth many billions of dollars of new extraction contracts for oil, gas, minerals, and so on.

Second, such new isolated urban nuclear power plants on barges potentially could be also a precursor to a similar multinational commercial oil, gas, mineral, fishing, and fish farming exploitation of the continental shores of Antarctica.

Third, the United States and Europe originally developed third-generation fast breeder reactors and the initial development of fourth-generation breeder reactors. But it is not yet clear whether other Asian or Western nations will follow the very large-scale joint ventures between the Chinese and Russian nuclear power fast breeder reactors of the third-generation and even fourth-generation reactor design models. Although it is known that the United States and India have entered a joint venture to develop fast breeder reactors, it is not known whether they or others will be exploiting the newest French and Chinese joint venture in very large-scale continual reuse and reprocessing of used nuclear material. This huge volume of reprocessing of used nuclear fuel does lend itself to fast breeder reactor's very high levels of production and at present, there are significant safety questions about the very large-scale continual reuse and reprocessing of nuclear spent fuel. This particularly relates to the absolute segregation of, and the absolute safety and the containment of, the specific amounts of spent nuclear fuel, the reused nuclear fuel, new uranium, new plutonium, and finally what is called "MOX fuel," a separate and exacting chemical mixture of three previous fuels.

Finally, as Dan Yurman points out:

> One key question is exactly how much MOX or combined spent uranium fuel, plus new uranium fuel, plus plutonium, China would be able to produce in that one decade. According to published data on French spent fuel reprocessing and MOX manufacturing, French calculations estimate that 800 tons of spent fuel would produce 8 tons of plutonium and 760 tons of uranium. The two would be combined with enriched uranium to produce MOX fuel equal to uranium fuel at 4.5 percent u235.[9]

In Chapter 15, we discuss hydropower plants.

---

[9] Yurman, "China New Reprocessing."

Hydropower Plants

Nothing is 100 percent; everything is a weighted probability.
—President Bill Clinton

Interesting investment opportunities are available in terms of operating hydropower plants. Hydro is the only carbon-neutral energy source that is economic. The best project opportunities involve low-impact hydro projects. Projects of this type have become accepted by nongovernmental organizations (NGOs) as a bona fide clean energy source. In some states, regulators are looking for ways to encourage the optimization of operating hydro plants. These hydropower plants can also qualify for renewable energy credits (RECs). As it is for wind and solar power projects, it is tough in the current market to make the economics work on the development of a new hydropower plant.

A UNIQUE RENEWABLE TECHNOLOGY

Globally, hydropower provides 16 percent of electricity, slightly more than nuclear power and closing in on natural gas, according to the London-based International Hydropower Association. In the United States, by contrast, hydropower now provides about 7 percent of electricity generation. All other renewable sources combined account for about 3 percent.[1] Norway has 12,000 megawatts (mW) of hydro that is more than 30 years old which runs on a merchant basis.

---

[1] Stephanie Simon, "Water Surge Hydropower, Once Shunned Because of Environmental Concerns, Is Making a Comeback," *Wall Street Journal*, September 13, 2010.

Older, operating hydropower plants with power purchase agreements that are indexed to avoided cost are good restructuring candidates. Since avoided cost is typically determined by natural gas in most markets, this tends to be a natural gas–fired power plant. Power projects that are based on the current forward curve are also facing revenue challenges. The low cost of natural gas is reducing the overall cost for electric power and reducing the cash produced by a hydropower plant. Some older hydro projects even have a price for power that is based on a discount to the forward curve for power. In certain states, electric utilities may have a tracking account that takes into account any front-loaded power pricing. Since this account could be rejected as an unsecured claim against the estate in a bankruptcy filing, it might be secured by a second lien to the project's debt. The fulcrum security for these projects will be the secured debt and not the equity.

Unlike wind and solar projects, most hydro projects have a relatively high yearly availability. A run-of-river project can have an average availability of 65 percent or higher over a 10-year period. These projects tend to have low head (e.g., less than 30 feet). They are located at dams built for flood control or navigation. There are 80,000 dams in the United States that don't have hydropower plants.[2] This can allow the hydro project to qualify for a higher-capacity payment revenue stream than a wind or solar power plant. Insurance companies like to invest in the debt of hydropower plants. The strong, long-term cash flows match the funding requirements of their required policy payouts. Insurance companies are willing to provide relatively high levels of leverage over a long term to hydro projects as opposed to wind and solar due to these issues.

The relatively high availability of hydro projects can complement wind and solar power plants. This situation may not occur during dry years. Some hydro projects also have low power output during the critical summer capacity period. Water flow to a facility may also be reduced by the local water supply authority or the Army Corp of Engineers for other reasons. If the plant has a firm energy or capacity obligation this can cause an issue.

There is also an extensive amount of information that can be learned about operating hydro plants on the Federal Energy Regulatory Commission's (FERC's) web site. The FERC grants each hydropower plant an operating license for an initial 50-year period. Before a license is issued the FERC takes comments from the local community and considers environmental issues such as the impact on local fish. After this period, the project can apply for an extension of its license or relicensing. Each hydropower plant has its own unique P number. The procedure is to enter the P number and the four-digit unit code (e.g., P-0000). The following FERC web site is

---

[2] Simon, "Hydropower, Once Shunned."

very useful for researching information on operating hydropower plants: http://elibrary-backup.ferc.gov/IDMWS/search/fercgensearch.asp.

HYDROPOWER AND RECS

Existing hydropower plants can have trouble qualifying for Tier 1 or any REC credits depending on the date that they were built. The typical position of a regulator is that a subsidy should not be given to an existing project. Their reasoning is that the project is already built and shouldn't require any additional subsidies. Operating hydro projects also don't qualify for the production tax credit (PTC) or investment tax credit (ITC). The extended development and permitting period for a hydro project also make it likely that any existing tax ITC and PTC credits may expire. This situation creates a challenge for hydropower projects in the current low natural gas and no CO_2 tax environment. The revenue stream from a strong REC credit stream is important to hit an acceptable internal rate of return from a hydro project.

In order to obtain low-impact certification, hydro projects have been turning to the Low Impact Hydro Institute (LIHI). In a number of power markets, it is necessary to become certified by LIHI to qualify for RECs. LIHI reviews each hydro project's FERC filings and makes sure that the project is in compliance with all of its permits. They are also concerned about the projects effect on local wildlife. The approach with current hydro technology is to achieve turbine efficiencies greater than 90 percent and fish passage survival greater than 96 percent.

Engineering Report

An outline for an engineering report that would be obtained when considering the acquisition of a typical operating hydro power plant would include the following issues:

1. Executive Summary (one-page)
   a. No deal-killers/Overall VG condition/NPV impact of recommendations
2. Introduction
   a. Purpose
   b. Scope
   c. Organization of report
3. Project Summaries
   a. Project description
   b. Condition assessment
   c. Part 12 issues

4. Regulatory Compliance
   a. Introduction
   b. FERC 101
   c. Project summaries
5. Energy Review
   a. Introduction (organize by river system, methodology)
   b. Summarize historical generation
   c. Review studies by other consultants
   d. Discuss any adjustments/changes
6. Going Forward Costs
   a. Introduction
   b. Cap Ex
   c. Op Ex
7. Conclusions and Recommendations
   a. Summarize:
      i. Study results
      ii. Significant findings
      iii. Recommendations
      iv. Roll-up operating data to real time data base
      v. Complete installation of new PLC operating systems
      vi. Implement specific measures:
         1. Project #1—install cathodic protection on penstocks, etc.

Engineering consultants also estimate replacement and operating and maintenance costs. The following text outlines a typical calculation for a medium sized hydro power plant.

Operations and Maintenance Costs

Using 200,000 mWh for annual generation, a $2.8 million budget represents 1.4 cts/kWh—which is in the range of industry experience. We've seen values anywhere from 0.5 Cts/kWh to 2 cts/kWh or more, depending on the specifics of each individual project. Major components of operations and maintenance (O&M) include staff labor, third-party services (contractors, regulatory compliance, instrumentation and control technicians, etc.), minor and major overhauls, consumables, property taxes, insurances, leases, and fees.

We have no information on taxes, insurance, and fees but we can share the following high-level planning numbers with you:

■ *Labor.* Being located at an Army Corps Lock and Dam facility, the powerhouse may have to be staffed 24/7. The minimum staff size to cover this is seven full-time employees (FTEs) plus a plant manager. A fully loaded rate with some overtime would be in the range of $80,000 per FTE. A good plant manager with incentives would be in the range of $150,000. A ballpark number would be $(7 \times \$80,000) + \$150,000 = \$710,000$.

- *Minor and major overhauls.* Following is an overhaul (OH) budget from a similar project on the Ohio River that we worked on in 2008. Minor OHs work out to approximately $25,000/year per unit ($70,000/3 + esc to 2010). Major OHs would be approximately $40,000/year per unit ($550,000/15 + esc). For two units, the annualized OH costs would be approximately $130,000 per year [2 × ($25,000 + $40,000)].
- *Third-party services.* Could be anywhere between $50,000 and $250,000 per year; say $150,000.
- *Consumables.* Allow $100,000
- *Subtotal:*

Labor	$710,000
OHs	$130,000
Third party	$150,000
Consumables	$100,000
$1,090,000	Say: $1.1 million

This would leave $1.7 million per year for property taxes, insurances, leases, and fees, which sounds reasonable—and perhaps a bit conservative—depending on specifics of the project.

Relicensing

Major issues regarding relicensing typically include fish passage; ESA (rare, threatened, and endangered species) and recreation. We do not anticipate any significant Part 12 issues (Dam Safety). We researched the FERC web site, relevant agency correspondence, and the license. Here's what we found:

- *Fish passage.* The dams are so far up in the watershed, and there are so many large dams downstream (on the Allegheny and the Ohio) that we do not anticipate any anadromous fish passage issues. A quick look suggests that nobody is pushing a big anadromous restoration plan. Also, because these are on locks, it is likely fish are moving up- and downstream now. There could be entrainment concerns, now or in the future, but I have not seen any agency nastygrams to suggest it is something of current concern.
- *ESA.* Potential issues would likely include mussels. It looks like there have been some mussel surveys in the area in the recent past. The issue here would be the potential for something to be found or listed between

TABLE 15.1   Typical Cash Flow for a Five-Year Relicensing

Year	5	4	3	2	1
Budget Estimate	$50,000	$100,000	$150,000	$150,000	$50,000

now and relicensing. We accounted for this in the relicensing cost estimate.

- *Recreation.* Recreation use at Army Corps Lock and Dam facilities is typically limited to day use (fishing access, picnicking, etc.). No significant issues are anticipated, but we included a modes allowance in the relicensing estimate.
- *Budget.* To be safe, we would put each facility at a $500,000 licensing cost in 2009 dollars. That would cover $150,000 in consultation and filing costs and $350,000 for studies and fieldwork. Given the size of the river, any studies will be expensive, and I would anticipate mussel surveys, maybe some fish entrainment studies, recreation studies, and maybe some water quality modeling (though dissolved oxygen does not appear to be an issue currently). Table 15.1 shows a typical cash flow for a five-year relicensing, and Tables 15.2 and 15.3 show various costs.

TABLE 15.2   Two 8m Tube Units at 17-Foot Head Rated @ 37 mW Total

No.	Item	$1,000s
1	Mobilization, demobilization	$8,645
2	Cellular cofferdam	$9,724
3	Intake structure	$24,003
4	Powerhouse	$27,292
5	Tailrace	$908
6	Access building	$2,704
7	Turbine and generator	$27,090
8	Substation	$5,854
9	Gates and miscellaneous mechanical	$17,082
10	Access roads and fishing pier	$2,600
11	Total direct costs	$125,902
12	Indirect costs	
a	Engineering	$6,295
b	Administration/Regulatory	$3,148
c	Subtotal:	$135,345
d	T/G contingencies (10%)	$3,294
e	Civil works contingencies (20%)	$18,592
13	Total estimated construction costs	$157,231
	$/kW	$4,249

TABLE 15.3 Adjustment to 14-ft Head

mW @ 17-ft Head	37
mW @ 14-ft Head	30.5
$/kW @ 17-ft Head	$5,160

TABLE 15.4

Project	Unit 1	Unit 2
Replacement cost	$5,160/kW	$4,250/kW

Based on our work at a similar two-unit project on the Ohio River, we developed the replacement costs shown in Table 15.4.

These values reflect only the direct construction costs. They do not include licensing costs, site acquisition costs, lenders' fees, and so on.

HYDROPOWER ECONOMICS

Hydro projects are optimized at the time of development and have to be optimized to take maximum advantage of current average annual river flows. This optimization can include new hydraulic power units, new governors, remote pond control, and movable forebay wall gate. This upgrade requirement is no different than for other operating power plants.

All projects can benefit from the sale of different products in different markets. In the case of the Marcus Hook power project, its capacity is sold to New York Independent System Operator (NYISO) Zone K (Long Island) via the Neptune transmission line and its energy is sold to the PJM market. The following text provides additional information on this project:

The Neptune cable that runs from New Jersey to Nassau County on Long Island can carry 660 megawatts (MW) of power, which is enough to meet the average electric demand of about 600,000 homes. LIPA customers saved over $20 million in the summer of 2007 by using the new Neptune cable to bring nearly 1.2 million megawatt hours (MWh) of low cost power to Long Island during the peak summer season in July, August and September when demand for electricity was highest. LIPA purchases of wholesale power from the PJM power market provided economic and reliability benefits for LIPA's customers in 2007 and are expected to continue for many years in the future.

In 2006 LIPA selected FPL for 685 MW of capacity from the Marcus Hook generating station in New Jersey. The Marcus Hook purchase is for capacity only with emergency energy, i.e., LIPA can purchase energy from it if both NYISO and PJM ISO declare a system emergency. Otherwise, the capacity is purchased as an unbundled market product used to meet on-Island and statewide capacity obligations set by NYISO. LIPA can and will use the cable to import energy purchased elsewhere, such as the spot market and PJM wholesale market. The Marcus Hook purchase is scheduled to begin on June 1, 2010. LIPA will continue to make energy purchases from a variety of market sources to import over the Neptune cable after the Marcus Hook capacity purchase begins . . . ."[3]

Pumped storage facilities pump water to a higher elevation in order to turn a turbine to produce electricity. The water is moved using cheaper power during off-peak periods. These facilities are typically located in the side of a mountain. During on-peak periods, the water is released to drive a hydro turbine. The concept is that since wind turbines produce nonfirm power, a pumped storage facility would act like a battery to store power. Unless the wind project is located next to the pumped storage facility, an additional investment in transmission will be required. In certain markets, wind power operates at night, which can integrate well with pumped storage. China has 2,200 pumped storage projects under construction.[4]

The challenge in the United States is that most capacity markets have a term of only three years. Pumped storage facilities are expensive and can easily exceed $2,000/kW.[5] It is also difficult to obtain local approvals and other permits for these facilities. As a result of the low price for natural gas, the intrinsic economics of pumped storage plants is negative. This is another example of inexpensive natural gas having a perverse incentive on not just renewable power but also storage. The round trip efficiency of a pumped storage plant is typically 70 percent.[6] Using the forward curve from the PJM market, the economics of a pumped storage plant as of April 1, 2011, would be as shown in Table 15.5.

These issues show that pumped storage projects suffer from a market failure.

---

[3] www.lipower.org/pdfs/company/projects/energyplan09/energyplan09-b.pdf.
[4] Simon, "Hydropower, Once Shunned."
[5] Mark Griffith, "Conquering Time: Understanding the Value of Pumped Storage," *Public Utilities Fortnightly*, October 28, 2008.
[6] Griffith, "Conquering Time."

TABLE 15.5 Simplified Pumped Storage Economics

	On Peak	Off Peak	70% Efficiency	Margin
2012	53.40	39.50	$39.50 \times 1.43 = 56.49$	−3.09
2013	55.40	41.25	$41.25 \times 1.43 = 58.99$	−3.59
2014	58.75	44.25	$44.25 \times 1.43 = 63.28$	−4.53
2015	62.25	47.25	$47.25 \times 1.43 = 67.57$	−5.32

SUMMARY

Hydro plants have a low environmental impact and can have a high capacity factor and they can produce power much more reliably than wind and solar power plants. However, they are expensive to construct and it can be difficult to obtain permits for new sites. But hydro power plants with damming can provide peaking power that, depending on transmission, can be combined with other renewable sources of energy.

In Chapter 16, we discuss geothermal power plants.

Geothermal Power Plants

Pity the planet, all joy gone from this sweet volcanic cone.
—Robert Lowell

Over many millions of years, the history of our Earth has been shaped by countless volcanic eruptions, which created thousands of mountains and mountain ranges and countless islands such as Hawaii, the Philippines, Iceland, and others. In Thera, a Greek island off the coast of Crete in the Mediterranean, a volcanic eruption exploded and destroyed the ancient Greek Minoan civilization. In Italy, Mount Vesuvius's volcanic eruption destroyed the ancient cities of Pompeii and Herculaneum.

According to the U.S. Department of Energy (DOE), under the earth's surface there is a layer of hot molten rock called magma, which is the source of geothermal energy. Today, this magma drives 8,900 megawatts (mW) in large-scale industrial power plants in 24 nations. The places in the world with the highest temperatures underground often have active young volcanoes that are sometimes called *hot spots*. These occur where the giant continental tectonic plates meet each other and the Earth's crust is thin, enabling heat, fire, or geysers to break through the Earth's surface. The world's largest numbers of hotspots with volcanoes are found in many places, but especially on the boundaries between these continental plates in what is known as the Pacific Rim on the "Ring of Fire." These hot spots occur in the Philippines, Japan, Alaska, California, Nevada, Mexico, and El Salvador.

STEAM TECHNOLOGY

The largest collection of hot spots in the world is in northern California, where dry steam spouts from many cracks in the Earth's rocky crust and continues every day. This location is called "the Geysers." The dry steam from many of these cracks in the Earth rises directly into turbines of many geothermal power plants that are placed on top of these geologic cracks. The dry steam drives the turbines, which directly drive electric generators that capture the geothermal energy, and transfers that power directly to the electric power station. Electricity from the Geysers is then distributed out to the West Coast high-voltage electricity power transmission system, which is a part of our interconnected national electric grid. The Geysers is owned by a major electric utility company called Calpine, which supplies most of the northern coast of California up to the border of Oregon. The Geysers plants use an evaporating water-cooling process to create a vacuum that pulls steam through the turbines more efficiently. However, this water-cooled process loses 60 to 80 percent of the steam into the air. It does not inject it back into the ground. Therefore, when the steam pressure declines, the rocks underneath remain very hot. The result is that 11 million gallons of water every day is treated separately and must be transported to the Geysers plants from a wide radius. In short, there are significant costs to balance against the free geothermal energy from the center of the earth.

For decades, this dry steam technology was the most common in use at all geothermal power plants around the world. However, the dry steam required the power plant to be built actually to sit on top of the crack in the earth where the dry steam escaped. This limited the amount of geothermal power that could be obtained from the earth, and so a second new technology was created. Boreholes were drilled into the earth in many nearby locations, where the geothermal liquid could be obtained. That geothermal liquid, which was at temperatures higher than 360 degrees Fahrenheit, could now be used to spray into a giant container holding a different liquid that was held at a much lower pressure than the geothermal fluid and thus cause the new fluid to instantly "flash steam." That flash steam could then drive a turbine, which could drive an electric generator. This new flash steam was useful in so many more locations than the number of dry steam locations directly on top of volcanic cracks that flash steam became the more successful technology, and soon it became the most widely used and the most successful geothermal power plant design in the entire world.

However, after a number of years, it was discovered that most geothermal regions have moderate-temperature water in them that is lower than the hottest geothermal fluid. Scientists found that energy could be extracted from these lower-temperature fluids and thereby developed what became known as a binary-cycle power plant due to the use of two different fluids. Both the geothermal fluid and a secondary (binary) fluid that has a much lower boiling point than water pass through a heat exchanger. Heat from the geothermal liquid causes the secondary fluid to flash into vapor, which then drives the turbines and then the electric generator. This binary-cycle power plant is a closed-loop system so that almost nothing escapes into the atmosphere.

Because moderate-temperature water is the much more widely found geothermal liquid resource across the world, it is widely forecast that the majority of new geothermal power plants will probably be using the binary-cycle power plant design.

Geothermal energy and geothermal electricity are now being scientifically investigated in a number of different ways in both government and university labs, as well as in companies across the globe. That is because the magma at the center of the earth plus millions of the very hot dry rocks provide a huge geothermal resource that is cheap, clean, and virtually unlimited once science creates new technologies to use them commercially. Across America, the DOE has spent millions of dollars in its geothermal technology development programs on the physics of magma in different forms and different conditions, as well as hot rocks, to sponsor labs at universities in research and development working with U.S. national laboratories and scientists, venture capitalists, and large companies in the geothermal energy field to specifically drive down the cost of geothermal power to three to five cents per kilowatt hour (kWh). If it were achieved, this three- to five-cent cost would be much more cost competitive with natural gas.

GEOTHERMAL PROJECT COSTS

Geothermal power plant projects are very expensive compared to natural gas–fired power plants. These projects involve drilling risk from project inception and during operations. Existing wells continue to need maintenance and may quit after a certain operating period. This situation requires the owner to drill new project wells.

According to the DOE's Energy Information Administration (EIA), a dual flash steam geothermal facility will have an overnight cost of $5,578/kW. The same source estimates that a binary geothermal facility will have an overnight cost of $4,141/kW. Unlike natural gas plants, geothermal facilities benefit from five-year modified accelerated cost recovery system (MACRS) depreciation, depletion, and an annual production tax credit or cash grant. (Please note: the MACRS is the current tax depreciation system in the United States. Under this system, the capitalized cost [basis] of tangible property is recovered over a specified life by annual deductions for depreciation. The lives are specified broadly in the Internal Revenue Code. The Internal Revenue Service (IRS) publishes detailed tables of lives by classes of assets. The deduction for depreciation is computed under one of two methods [declining balance switching to straight line or straight line] at the election of the taxpayer, with limitations.[1])

With the low price of natural gas reducing the price of power, even these various tax benefits make the overall economics of geothermal projects difficult. Although America has more geothermal energy capacity (3,000 mW) than any other nation, U.S. geothermal energy power plants are located in only eight states. Three other nations of the world obtain over 25 percent of their entire energy capacity from geothermal power plants. These three nations with many hotspots are Iceland, the Philippines, and El Salvador. The geologic structure and moonlike landscape of Iceland, with steam rising from cracks in the rock everywhere, is particularly striking.

Eighty percent of U.S. geothermal power capacity is produced in California. Over 40 geothermal plants produce 5 percent of that state's electricity.

HYDROTHERMAL POWER SYSTEMS

The earliest known commercial uses of geothermal energy systems dating back to the ancient Egyptian, Chinese, and Roman Empires were for medical spas and swimming pools all the way from central China to Egypt to Spain, and in the north to Great Britain and Hungary. A number of those medical spas were used for many hundreds of years, and some are still in use today.

The most common commercial uses of low-temperature geothermal energy today are in greenhouses and fish farms. Most greenhouse operators estimate that using geothermal resources instead of traditional energy sources saves 80 percent of fuel costs. They also estimate that they save 5 to 8 percent of total operating costs. In Holland, the growing of flowers and fruits and vegetables in water in hydroponic glass "hothouses" is the most efficient and least expensive agriculture. The temperature remains constant. Each crop matures more quickly and, at the same time, with virtually no disease, blights, or wind or weather deformities. The same fertilizer stays in the water for a number of crops and does not need to be thrown out with each crop.

GROUND-SOURCE HEAT PUMPS

Throughout the entire year, the temperature of the Earth, eight feet beneath the surface, remains relatively constant, at between 50 and 60 degrees Fahrenheit, depending on the latitude. A geothermal heat pump or ground-source heat pump extracts underground heat in the winter for heating houses, office

---

[1] See IRS Publication 946 for a 120-page guide to MACRS.

buildings, schools, prisons, or hospitals, and then transfers the heat back into the ground in the summer for cooling. Ground-source heat pumps are extremely well matched architecturally to underfloor heating and baseboard radiator systems, underground garages, work rooms, and finished basements for living quarters.

In the United States, it is vital to realize that ground-source heating and cooling is far more efficient than using electric or oil heating and cooling. In fact, these natural thermal systems move up to five times the energy they use and significantly heat and cool homes at the least expense over their useful lifetime, which can be several decades. The Economic and Stimulus Emergency Act of 2008 includes an eight-year extension to 2016 of the 30 percent investment tax credit for geothermal heat pumps, with no upper limit to all home installations of "Energy Star." The simplest geothermal improvement is to lay a 25- to 40-foot pipe eight feet below ground and then feed it into your home.

Although the heat pump was first described by Lord Kelvin in 1853 and developed by Peter Ritter von Rittinger in 1855, it was not until 1946 that the first successful commercial project was installed in the Commonwealth Building in Portland, Oregon. Sweden popularized the commercial heat pump for heating and cooling, and there are now several million heat pump units installed worldwide. Open-loop heat pump systems were by far the most popular until 1979, when the closed-loop polybutylene pipe proved to be economic, and they have become more popular with a mixture of water and antifreeze. Open-loop heat pumps use natural ground water.

There are several different designs of geothermal heat pumps or ground heat pumps, starting with the direct-exchange pump, which is the simplest, easiest, and cheapest and uses a loop of copper pipe buried underground. The copper tube's thermal conductivity is higher than plastic pipe, which contributes to its greater efficiency; although it uses more refrigerant than closed-loop water systems, the direct exchange requires only 15 to 30 percent the length of tubing and half the diameter of drilled holes. Braised copper tubing is required; otherwise, the gas in the antifreeze can leak out.

Closed-loop systems need a heat exchanger between the refrigerant loop and the water loop and electric pumps in both loops. Closed-loop systems have lower efficiency than direct-exchange systems, so they require much longer and larger pipe in the ground; therefore, their excavation and installation costs are higher. Closed-loop pipes can be installed vertically deep in the ground five or six feet apart. A hole is bored 75 to 500 feet deep for large buildings in cities if there is a tight restriction on available land for the heat pump system. Closed-loop heat pump systems are also laid horizontally as loop fields or slinky loops in trenches that are deeper than the frost line. The cost of excavation for a horizontal looped heat pump field is far less than half the cost of a vertical field and is by far the most commonly built. Slinky tube fields are used if there is not space for a horizontal trench heat pump field. Slinky coils lay on top of each other in three rows across the trench, so the excavated trench is much shorter than a normal horizontal ground loop heat field and cheaper. Horizontal directional heat pump fields can be built under driveways or gardens or outbuildings if land is limited or aesthetic precision demands it.

STANDING COLUMN WELLS

A multiple standing column well system is the largest form of ground pump that can support a city or town. There are many successful multiple standing column wells in the five boroughs of New York City and also throughout the states in New England. A standing column ground-source well system has heat storage benefits where heat is rejected from the building and the temperature in the well is raised during the summer months, which can then be harvested for heating during the winter months, thus increasing the efficiency of the heat pump system. However, the sizing of the standing column well system is critical as it relates to the heat gain or heat loss of the existing building or the town or city. Because the heat exchange is actually with the bedrock, it uses water as the transfer medium. However, if there is adequate water production, then the thermal capacity of the well system can be enhanced by discharging a small percentage of system flow during the peak summer. Since this is essentially a water pumping system, standing column well design requires critical considerations to obtain peak operating efficiency. Should a standing column well design be misapplied or leave out critical shutoff valves, for example, the result could be an extreme loss in efficiency and thereby cause operational cost to be much higher.

ENHANCED GEOTHERMAL SYSTEMS

Enhanced geothermal systems (EGSs) are being installed by the Departments of Energy in many nations. According to the U.S. Department of Energy:

> *EGSs in the United States, Australia, France, Germany and Japan. Enhanced Geothermal Systems are also installed by many corporations, and by many Universities, and by Venture Capital firms including Google to enable capturing the heat in dry areas called Hot Rock Reservoirs typically at greater depths below the earth's surface than conventional sources, are first broken up by pumping high pressure water through them. The plants then pump more*

water through the broken hot rocks where it heats up, returns to the surface as steam, and powers turbines to generate electricity. Ultimately, their water is returned to the reservoir through injection wells to complete the circulation loop. Enhanced geothermal systems that use a closed loop binary cycle, release no fluids or heat-trapping emissions other than water vapor which may be used for cooling.[2]

One key risk of EGSs is that they can cause increased seismic activity, which induces many small earthquakes similar to those that result from extensive hydraulic fracturing and drilling used to increase the rate of oil and gas production from huge shale rock formations. EGSs and increasing carbon sequestration capture and storage, in deep saline aquifers activity, can be dangerous to surrounding populations, especially if built near major geologic fault lines. These must be checked in advance and then must be monitored regularly.

Coproduction of Geothermal Electricity in Oil and Gas Wells

Co-production of geothermal electricity in oil and gas wells is a major future growth market. In the oil and gas fields, which are producing very well already, there is virtually always the highest probability for electricity production simultaneously. An MIT study forecasts that the United States has the potential to develop 44,000 mW of geothermal power by 2050, primarily in the Southeast and the Southern Plains states, which would supply 10 percent of the base-load electricity for all America.

All these dynamic new geothermal developments are funded by major DOE grants for 10 large research-and-development demonstration projects to prove the feasibility of enhanced geothermal systems technologies and also to prove the viability of low-temperature geothermal projects.

Magma from the center of the earth and hot, dry rocks will provide almost unlimited energy that is clean and cheap, as soon as we develop new technology to use them safely. Until then, moderate-temperature sites running binary-cycle power plants are going to be the most common of all geothermal plants.

DIRECT USE OF GEOTHERMAL ENERGY

Geothermal energy is widely predicted to be heavily funded by many governments because it is among the lowest-cost heating and cooling systems and because it has been found to be available all around the world.

---

[2] http://energy.gov/.

Geothermal temperatures of 50 to 60 degrees can be widely used for heating homes, offices, commercial greenhouses, fish farms, gold-mining operations, and a variety of special applications in addition. Spent fluids from geothermal electric plants can be used subsequently for direct-use applications called *cascaded* operations. Savings can be as much as 80 percent under the cost of fossil fuels. In addition, geothermal energy usually has few, if any, pollutants, emissions, or toxic residues. The primary uses of direct geothermal heat are in district and space heating. A survey of 10 western states identified more than 9,000 thermal wells and springs, plus more than 900 low- to moderate-temperature geothermal resource areas and hundreds of direct-use areas.

"Dry heat power is a specialized form of geothermal energy that escapes from hundreds of cracks in the Earth's surface. In order to fully exploit dry heat geothermal power, many very deep holes are drilled in the rock or the ground, and heat pumps are inserted down the holes to drive the heat up to the Earth's surface, where it is captured to be used directly in new commercial operations or municipal heating or cooling operations. In addition to the United States, Germany is substantially increasing government feed-in tariffs for geothermal energy development and distribution.

Although geothermal energy is widely available globally, it does require the coexistence of very substantial amounts of heat, fluids, and permeability in reservoirs. If hydrothermal reservoirs do not exist, these can be engineered or man-made by exploiting hot rocks deep in the ground for commercial use. This alternative is widely known as a form of hot rock enhanced geothermal systems.

There are normally five steps in the decision process for developing Enhanced Geothermal Systems and they include:

1. Find a site where hot rocks exist.
2. Create the reservoir.
3. Complete a well-field.
4. Operate the reservoir.
5. Operate the facility.

Hot rock enhanced geothermal reservoirs require drilling wells down into hot rocks and fracturing the rock sufficiently to enable water to flow between the wells. The water flows along what are called permeable pathways picking up heat, and finally exit the reservoir production wells to complete the circulation loop. (Note that if the plant uses a closed-loop binary cycle to generate electricity, none of fluids vent into the atmosphere.)

The adequacy of all these technologies for creating an EGS has been determined for both near-term and long-term applications. It exists.

However, to fully achieve large-scale (100,000 mW) use of cost-competitive geothermal energy, significant advances are needed in the site characterization, reservoir creation, well-field development and completion, and system operation, as well as improvement in drilling and power conversion technologies. Technology innovations and improvements will also support long-term ongoing development and expansion of the hydrothermal industry. In order to realize the promise of EGS as an economic national resource, we have to create and sustain each reservoir over the economic life of each of the EGS projects."[3]

SUMMARY

For a renewable power technology, geothermal power plants have high availability and have to also compete against shale gas. There are many other uses being found for geothermal technologies such as food-processing facilities, gold-mining operations, physical therapy facilities, injection wells, and storage ponds.

Enhanced geothermal technologies are forecast to be a significant growth industry worldwide for the next two to six decades because scientists know that the geothermal layer of molten magma is larger in total than all other energies on Earth. However, the technologies we need in order to handle and manage that magma under fail-safe controls are one of the great challenges the world scientists and engineers face.

Chapter 17 discusses energy efficiency.

---

[3] http://energy.gov/.

Energy Efficiency and Smart Grid

With demand side management, it may never be necessary to build another power plant.
—Amory Lovins, Rocky Mountain Institute

The cheapest and cleanest source of power comes from permanently reducing or managing power demand during times of shortage. Energy efficiency is defined as a permanent change in the use of energy. Demand response is a temporary change in the use of power.

DEMAND-SIDE MANAGEMENT

Often referred to as demand-side management (DSM) or negawatts, and much like renewable power, this "source" has no emissions. The smart grid, advanced metering, and smart appliances allow for the expansion of DSM. Some investors feel that the smart grid could have more impact than the Internet. The future potential for DSM is also partly determined by the low price of natural gas and the cost of emissions such as carbon dioxide (CO_2). Utilities and independent power producers also have to meet cyber security regulations. This can stop the installation of smart meters. There is a concern that cheap natural gas could make the United States complacent on reducing the use of energy. DSM also faces the challenge that it might not be available to the utility when called on. Similar to wind and solar power plants, DSM is not a seven-day-a-week, 24-hour-a-day resource. As stated in *Bloomberg Markets:*

> Utilities and state regulators are loath to make sweeping changes to this system, especially after California's disastrous experiment in

deregulation that enabled Enron Corp. energy traders to create artificial shortages and rolling blackouts in 2000 and 2001.[1]

Utilities in New Jersey can't install smart meters because the New Jersey Board of Public Utilities will not currently approve them. There is also a case in front of the Maryland Public Service Commission by Baltimore Gas and Electric. A large challenge to this source is that most residential and commercial consumers don't see the real-time price that they pay for power. There is a joke in the utility business that states that if Thomas Edison returned to life, he would feel that nothing has changed in the electric utility industry. If Alexander Graham Bell returned to life, he would be amazed by what has changed in the telecom industry. In the current market, utilities don't know if service is out to an individual house. The individual has to call the utility. This results in their not having an incentive to reduce their electricity consumption. All customers are also not currently charged for transmission congestion and environmental degradation. There is also a landlord-tenant challenge in some apartment buildings. This situation results in the landlord's supplying the large, power-consuming appliances, while the tenant pays the electric bill. This reduces the incentive for the landlord to install more efficient appliances.

In the past, regulated utilities earned only on their rate base when they built new power plants or transmission lines. In order to encourage DSM, rate bases have been "decoupled" so that regulated utilities can now earn on DSM investments. This is referred to as performance-based rates or rate decoupling. This is favored by some over net metering since a utility might not be paying enough for the power it obtains from its customers. A business can increase its profits by either increasing revenue or reducing expenses. As a comparison, ConEd spends $1.2 billion per year on infrastructure.[2] One way to achieve demand response is by curtailing the central air conditioning or water heating in private homes. It is not uncommon for these controls to be overridden by the individual homeowners. This results in no loss of demand for the utility. A power plant would have to pay liquidated damages if it failed to deliver power. A penalty of this type is not typically required of a residential user of power. A permanent reduction in load could be achieved by converting an electric water heater to natural gas.

It is important to remember that customers want only what the energy provides. They really want cold beer and hot showers. However, behavior

---

[1] Edward Robinson, "EnerNOC Returns 260% from Lowering Lights in 2009's Power Grid," *Bloomberg Markets*, August 14, 2009.
[2] Consolidated Edison, "Steam Long Range Plan, 2010–2030" (December 2010), presentation from March 24, 2011.

change is often underestimated by the electric industry to implement DSM. Customers can change behavior when given information. Consumers want HBO, so they are willing to pay for cable. There are times during off-peak periods when the price for power is negative. This is due to the fact that wind projects can produce a large amount of their power during off-peak times. This is "cheap" power that could be tracked by the smart grid and sent to storage devices or electric vehicles. As noted elsewhere in the book, batteries have a very limited storage capability. One venture capital investor has stated that even with a 10-year fund life, this is too short a period to invest in electric vehicles. Subject to limits on transmission and distribution, the smart grid could also be a new provider of ancillary services and capacity.

Rewarding Efficiency

Con Edison has offered rebates to residential customers in the past who purchased more efficient window air conditioners. It is important to point out that a regulated utility will have a lower cost of capital and hence a lower investment hurdle rate than most of its customers. As a result, it could make sense for the utility to finance the DSM equipment or smart meter cost for its customers or offer a rebate to the customer that can reduce its electric load. This cost for the utility would be placed in its rate base, and it would be allowed to earn a return just like it could on an investment in a new power plant. An expenditure of this type is very cheap, quick capacity without the expense and risk of developing a new power plant. The utility has the obligation to serve, and as long as it can make the used and useful argument with its regulators, it has no electricity price or volume risk.

In a commercial application, an office building could agree to reduce its air conditioning load and send the employees home. A commercial building may also have a backup generator that could be operated during times of peak demand. Local air emissions regulations may limit this generator from operating or may limit the hours that it can operate. The problem would be if they went home and turned on their home window/wall air conditioning units. In fact, there are over 6.1 million window/wall units in New York City out of a total of 30 million in the entire United States.[3] Some of these air conditioners are located in master metered buildings, where individual residents don't see the price for power.

As compared to a residential application, there are a number of levers that can be pulled by an industrial or commercial electric power user to reduce electric load. This includes heating, ventilation, and air conditioning

---

[3] ConEd, "Steam Long Range Plan."

(HVAC), lighting, electric chillers (air conditioning), elevator banks, and data centers. The concept is that if enough buildings in a city area can reduce their electric load, it can act like a single power plant during times of peak power.

EnerNOC claims that is has 5,300-mW of capacity under its control from demand response as of December 31, 2010. This is equivalent to a number of large coal, natural, and/or nuclear power plants. As mentioned earlier, the constraint is that this resource is located in a number of areas and is available only during limited times. However, electric loads can also be aggregated in one particular area with no or limited transmission loss. An apartment building could potentially shut down its hall lights, some of its elevators, and its laundry room.

Unlike a power plant, there is no emission profile from reducing consumer and industrial load. This is especially true since the power plant that DSM typically competes with is a gas turbine engine peaking power plant that might have limited emission controls. It is expensive to stop and start a gas turbine engine. There is also a time limit on how quickly a gas turbine engine can be started from a cold condition. If a building started a standby diesel engine, there could actually be an increase in emissions, depending on the sulfur content of the fuel that is used and the emission controls on the diesel engine.

The best bang for the buck for DSM will come from commercial and industrial users of power. Some utility executives have become "smart grid" skeptics about the consumer DSM market. They feel that the amount of demand that can be shed from an individual apartment or house is not material and offers no real economic benefits. They don't see the business case for switching off the washing machine and upsetting consumers. Their feeling is that the smart grid is only a lobbying effort from information technology companies. However, the current installed base of meters provides only one-way communication, doesn't display real-time pricing, doesn't notify the utility of any power outage, and provides no real-time customer power consumption data.

ADVANCED METER INFRASTRUCTURE

The cost to install advanced meter infrastructure (AMI) is high, and regulated utilities are always concerned about increasing their rates and the "used and useful concerns of regulators." Both regulators and utilities are concerned about AMI technology becoming obsolete once it is installed. They are also concerned that consumers will accidentally continue to buy power during on-peak times and complain about high prices for power. Residental consumers have a limited ability to reduce or reschedule large amounts of electric load. Meters are expensive to change once they are installed. The installation on AMI becomes a simple net present value (NPV) calculation. The NPV calculation for this investment easily can be negative if only a small amount of load is reduced by the consumer. The electric bill is frequently the least expensive consumer utility bill.

There is a battle going on how AMI meters will communicate with the utility and/or independent system operator (ISO)/regional transmission organization (RTO) head office. Some technologies such as Silver Spring will be based on using radio waves. The concern is that this communication protocol may not be robust enough. Users of home wireless modems (WiFi) can experience problems with their Internet connection when a microwave oven is turned on. WiFi technology is based on radio waves. Radio waves can be especially problematic in city locations, where there are a lot of conflicting signals. Other smart meter developers are focusing on using WiMax or are teaming with cell phone carriers to send data over their networks. Supports of WiMax argue that "Its massive bandwidth and secure, standardized spectrum will enable utilities to cope with the possible huge flows of data from electric vehicles, rooftop solar panels and other green-energy technologies in the next decade."[4]

The cost to participate in demand response programs is rising due to ever-tightening rules by both ISOs and regulated utilities. These costs include the direct costs of aggregating demand response, the required profit margin for aggregators, the indirect cost associated with consumer inconvenience, and recently tightened penalties for nonperformance. Pure-play electricity aggregators such as EnerNOC are facing competition from suppliers of natural gas who can also offer demand response services. These suppliers already have access to industrial and commercial customers and can cross-subsidize demand response services with the sale of natural gas.

INCREASING ENERGY NEEDS

As mentioned elsewhere in this book, it is becoming more and more difficult to site new power plants. In addition to numerous locations being declared nonattainment for criteria air pollution such as nitrous oxide (NOx), they can also be designated as an environmental justice area. Each individual state Department of Environmental Protection will take these issues into account before granting a new air permit. Nongovernmental organizations (NGOs) such as the Environmental Defense Fund, the National Resources Defense Council, and the Sierra Club will also opine on the siting of new

---

[4] Robinson, "EnerNOC Returns 260% from Lowering Lights in 2009's Power Grid."

power plants. The challenge for a new plant developer is that there might not be any due process with NGO groups.

It is not just air conditioners that have high power demand. The new 3D television technology requires 30 percent more power than older TVs. Homeowners will agree to have their air conditioning curtailed by their local utility but not their TV. Consumers would also not accept being forced to spontaneously shut off their personal computers or to stop charging their cell phones. A cable TV technician once told one of the authors that consumers can live without their TV service but not their Internet. Residential customers will not want their washing machines stopping before they finish their cycles. This load can be partly offset by new buildings that are designed to be energy efficient from initial design. The changes that can be made on existing buildings are more limited.

If plug-in hybrid electric vehicles (PHEVs) become more widespread in the future, utilities will want to manage when they are charged. They will not want cars to be charged at 4 p.m. The preferred time would be between 2 a.m. and 7 a.m. As pointed out elsewhere in this book, current battery storage technology is very limited. This fact will ultimately limit the adoption of vehicles of this type.

In New York, Con Edison has designated its demand response season from May to October. This time period is coincident with its summer peak period. Utilities, ISOs, and RTOs still face customer barriers to the introduction of DSM and energy efficiency programs. According to ConEd, these barriers include ignorance, fear, confusion, and economics.

At the present time, utilities are seeing a 2.5-hour response time for demand response resources and a 70 percent performance factor. This translates to 70 mW from every 100 mW of contracted load. Performance levels of this type are not competitive with a gas turbine engine. The future approach will be automated demand response (ADR). This would have a 4- to 10-second response and a 90 percent performance factor. The concern is that this supply of power could still be overridden by the consumer that had "offered" this electricity supply to the local utility. ADR also remains untested in the current marketplace.

A Federal Energy Regulatory Commission (FERC) ruling dated March 15, 2011, stated that demand reduction should be paid the same locational marginal price (LMP) as a traditional generator would if it were selected for dispatch. Prior to this FERC ruling, PJM had attempted to limit the amount of megawatt hours that could be claimed by demand response. The following comment from EnerNOC further clarifies this issue:

EnerNOC firmly believes that end-users participating in demand response should be compensated for actual, verified load reductions provided to the grid, and this should not be limited by what their load happened to be in the previous year. The PJM and Market Monitor position is that if a customer reduces its real-time demand from 25 megawatts (MW) to 5 MW in response to a system emergency, but the customer's peak demand from the previous year was only 10 MW, then none of the load drop from 25 MW down to 10 MW should be compensated. This is impractical and unfair," said David Brewster, President of EnerNOC. *"The purpose of this filing is to enable EnerNOC to continue to manage demand response resources in accordance with existing market rules and established practices.*[5]

SUMMARY

The smart grid has a large future potential and also has to compete against shale gas. Like renewables and coal plants, demand side management faces tough competition from inexpensive shale gas. It is even difficult to make capital expenditures on more efficient lighting when the price for power is low due to inexpensive natural gas. Except for large users of power, it is difficult to shift meaningful amounts of load to off-peak periods. The expiration of the 1603 cash grant for renewable power projects may shift more interest to investing in DSM opportunities.

---

[5] EnerNOC filing from February 23, 2011.

Conclusion

Because huge volumes of shale natural gas have been discovered across much of America at currently very cheap costs, it is widely forecast that there will continue to be a glut of shale gas in the U.S. market. This cheap natural gas supply has made it very difficult for any other power technology to compete on price or to be economically viable. In the majority of cases of renewable energy projects, there is a very real question of whether they would have had better prospects if shale gas had never been discovered.

For the past 10 years, new shale natural gas deposits have reportedly been discovered in France, Germany, Poland, China, and other nations. However, because most other nations lack U.S. natural gas volumes, very specialized equipment, and expert teams for hydraulic fracturing for natural gas and natural gas liquids, they were not yet ready to become serious competitors to the United States in horizontal drilling, pre-perforated pipes, and advanced explosives and doing comprehensive scientific forecasts of total volumes of potential natural gas.

For years, the standard method of hydraulic fracking of a site took one or two weeks to first secure the actual site and determine the path of the drilling and to measure the pipe distances and perforate the top and bottom of specific pipe lengths that would be blocked off at a fixed end point. This was in order to separately explode one relatively short section of the natural gas field, extract the gas from it, and then move on to the next section of pipe each day until by the end of a week or two, the separate sections of pipe had been sequentially exploded.

However, today, this new natural gas hydraulic fracking advanced technology equipment race has become even more intense. That is especially because the most advanced horizontal drilling equipment piping today is pre-perforated above and below the pipe with all the explosive charges preset along the entire pipe. The multiple explosive charges are preset in order to complete the natural gas pipe explosions in three, two, or one giant explosion all in only one or two days, instead of much smaller explosions stretched over one or two weeks. There may be a potential increase in risks of pipe cracks.

As practiced now, natural shale gas fracking puts ready money profits into specific land owners' pockets. It is not yet clear whether there is a permanent economic improvement or a decline in the economy of many rural communities where natural gas fracking has taken place.

The *New York Times* front-page lead stories on both June 26 and 27, 2011,[1] stated that while natural gas was discovered across the world, the volume of shale gas from specific shale wells was unpredictable. This fact was widely known, as was the fact that some wells remained financially productive after five years, while many other wells had sharply declined, and some shale wells had petered out within a decade. The *New York Times*'s implication was that natural gas fracking was not the long-term oil drilling experience of Texas oil barons. Yet, a key point that was not stressed was that giant new natural shale gas fields are being discovered every month in different parts of the United States as well as simultaneously in different nations of the world.

The *New York Times* also reported that after 15 years, shale natural gas production from a number of these same hydraulic wells had so dramatically declined that some wells were operating at steep financial losses. At some smaller shale fields there was steep decline of gas production over time. The *New York Times* also stated, "There is undoubtedly a vast amount of gas in the [giant shale gas] formations. The question remains how affordably can it be extracted?" However, this evidence of steep decline in shale gas was not obtained from the Marcellus Shale Gas Formation, [which is the richest shale gas formation in the United States] but instead, from certain other southern and western much smaller shale gas rock formations.

This divergence of scientific results from shale gas production in different locations has led to two entirely different financial forecasts. In the first, the United States has an increasing glut of shale gas, which is leading to ever-larger investments both by small, independent companies and by giant energy conglomerates like ExxonMobil, Royal Dutch Shell and British Petroleum, Conoco Philips, Texaco, and Halliburton.

This glut of cheap shale gas coming onto the market now has appeared to have undermined already planned national and state funding, plus corporate funding of previously planned alternative investments in renewable power plants. This includes solar power plants, wind power plants, biomass energy plants, new run-of-river hydropower plants, and even combined-

---

[1] Ian Urbine, "Drilling Down: Insiders Sound an Alarm Amid a Natural Gas Rush," *New York Times*, June 26, 2011, A1; and "Behind Veneer, Doubt on Future of Natural Gas," *New York Times*, June 27, 2011, A1.

cycle energy plants, nuclear power plants, and geothermal power plants. In the case of each of the renewable energy plants, as well as the traditional power plants, all of them were several times more expensive than the natural gas-fired power plant alternative.

Unless abundant scientific and irrefutable clinical medical evidence is soon brought forward to prove that hydraulic fracturing of natural gas is a proven carcinogen or other type of severe health hazard, the shale gas industry lobbyists and corporate laboratories will claim that two decades of previously discovered health dangers from hydraulic fracturing have been carefully studied. They will say all of those toxic or dangerous chemicals previously found are now being banned or remediated or rigidly constrained in order to protect public health.

They will attest that none of the prior dangerous chemicals are now in the "multiple chemicals cocktail" that is added to the sand and millions of gallons of water that are pumped at very intense pressure into the shale drilling pipes. The entire shale gas industry has been permitted to do hydraulic fracturing in many states, and it has so intensely lobbied the U.S. Senate and House of Representatives, as well as many state capitals and states' legislators, based on the fact that shale natural gas is today the only major, high-volume, domestically produced, U.S. substitute for OPEC and other foreign nations' oil, that they anticipate governments at all levels in the United States will virtually have to grant widespread permission to do hydraulic fracturing. The drilling corporations have now finally agreed to publicly print a list of all the chemicals and other items used at each specific well drilled.

Scientists and technologists at auto manufacturers like Ford, GM, Volkswagen, and Toyota have already carefully analyzed shale gas [by converting "gas to liquids", and some have already demonstrated that natural gas–powered cars are commercially viable—the Honda Civic natural gas–powered cars are already in production and are being sold. Likewise, aerospace companies have already done extensive tests on shale gas by converting "gas to liquids" to determine if it can be converted to a safe and acceptable substitute for jet fuel on major national carrier jet fleets. The major national and international airlines are very interested in the scientific proof that this natural gas based converted jet fuel is totally safe, can be manufactured in huge volume, and can become a major commercial success. This should be possible since natural gas turbine engines used in commercial airlines are also used in land-based power projects. That is because today's sky-high cost of jet fuel and its continuing escalating costs, plus jet fuel surcharges on passenger ticket prices and air cargo fleets' jet fuel prices, have eaten into world airlines profits. This could be expected to result in a major savings for airlines and also potentially for airline passengers and air cargo shipping firms if those savings were passed along.

Less expensive natural gas converted to liquid fuels as substitutes for jet fuel for jets and gasoline for cars could represent significant cost savings for the American consumer. This is clear because of the current universal jet fuel surcharge plus extra taxes on gas for cars based on "9/11 charges" and huge extra taxes on gasoline because of the huge negative volatility and uncertainty of foreign oil prices due to the continuing Iraq, Afghanistan, Libyan, and other continuing civil strife, civil wars, or religious or tribal conflicts or potential foreign military conflicts that might impact the daily price of oil and gasoline and jet fuel.

There is no doubt that a number of these cost reductions for airlines and cars could potentially or theoretically now be available for funding new renewable energies. However, it is clear that it is highly probable that corporate, federal U.S. government, and/or state funding would still be required in order to achieve commercial success in any renewable energy project.

It must also be remembered that for decades before today's shale natural gas was discovered, there were world markets for natural gas and liquid natural gas (LNG). It must also be realized that while natural shale gas is currently available at $4/MMBtu, there are Mideast nations and others where natural gas is plentiful and can be produced at only $2/MMBtu. Thus, the U.S. shale natural gas does not have the lowest global price and so could be looked at as a trapped basin that could keep future prices low. LNG export from the United States could be limited due to high cost, regulatory approvals, the relatively higher cost of its indigenous gas and the steep production decline curve of shale gas wells.

There are certain states, as well as certain foreign nations, that have already banned hydraulic fracturing of shale gas, shale gas liquids, and shale oil.

A number of the financial and economic consequences of U.S. states suddenly banning hydraulic fracturing of natural gas could be the exact reverse of the benefits that America would reap if hydraulic fracturing were to continue to be encouraged.

The United States may well be forced into a position where it has to compete against a variety of foreign national and consortia green energy investments run by foreign developers or foreign technologies. China is willing to fund new green energy prototypes and innovations to potentially achieve first-mover critical advances. China could be in a better technologic and scientific position to set newer renewable green energy industry-wide standards.

WHERE DO WE STAND TODAY IN TERMS OF RENEWABLE ENERGY?

To date, most wind and solar energy plants have been small scale. They are not able to achieve the required economies of scale.

There are a number of proposed large solar plants that recently won huge loan guarantees from the Department of Energy.

They required huge tracts of land on which to site hundreds of solar mirrors and tall solar towers, and some have large molten salt tanks to store the heat of the sun during the day to be used when the sun is not shining. These are extraordinarily expensive.

Similarly, original wind energy farms were often small scale. The next frontier for wind development is offshore and also are very expensive and very difficult to site.

Renewable power continues to be dependent on energy storage. As we've stated already in this book, renewable power is not a 24/7 resource.

Large amounts of additional research dollars are required to develop the car battery storage technologies, and new mass volume pumped air storage and huge solar power plants using molten salt tanks are major required improvements in U.S. energy development.

However, in both of these classic examples of renewable energy, where the highest total wind power and total solar power on earth are the greatest are most frequently located long distances from the major U.S. electricity grids. The high-tension electric transmission lines, and long distances from cities or major industrial centers where all that large amount of energy could best be put to its most efficient productive work.

Therefore, many new electric transmission technologies still must be further developed scientifically in order to put all these new advanced sources of renewable energy onto the national and international grids, where high-tech electricity transmission systems could now connect them to the industrial centers and cities.

Finally, following Japan's triple "meltdown" of three nuclear power plants and Germany and Switzerland's decisions to end their nuclear power plants, we see China, Russia, India, and South Korea making the greatest advances in nuclear energy growth. In terms of masses of new nuclear power plants, which their national governments are not only willing to fund to develop and build but also to buy in large volumes of new order flow. These nations are spearheading the fastest growth in reprocessing of their nations' existing nuclear fuel and thereby pushing the envelope of third- and fourth-generation nuclear power plants for their own use and for power plant sales abroad. It remains to be seen whether a new nuclear power reactor will be built in the United States after Japan and because of the glut of natural gas.

A "dash to gas" could create a dangerous dependency for the U.S. energy market. Only the U.S. government can fund and take on the risk of new energy technology.

As discussed throughout the book, renewable power projects have to compete against natural gas–fired power plants. The following term sheet (Table A.1) was filed with the Maryland Public Service Commission for a proposed 640-mW gas turbine combined-cycle power plant (GTCC). It provides a handy reference on GTCC performance and key power and natural gas market terms.

TABLE A.1 St. Charles Long-Term Contract Terms

Seller	CPV Maryland, LLC ("Seller")
Buyer	Buyer ("Buyer")
Products	Unit Contingent Physical Capacity
	Unit Contingent Energy Tolling (financially settled)
Contract Term	20 years starting at COD (expected Q2 2012)
	Note: Actual COD will be dependent on timing of execution of PPA with Buyer and other key transaction agreements.
Collateral	Seller will provide security, in a form to be mutually agreed upon, based on the following:
	$5/kW at contract execution
	$50/kW at Financial Closing
	$25/kW at Commercial Operations Date and reduced over the contract's life.
Project Milestones	Assuming that definitive agreements are executed, Seller shall guarantee the following Construction Milestones:
	Financial Closing by Q2 2010
	Delivery of Major Equipment to site by Q2 2011
	Commercial Operations Date by Q2 2012
	Failure to achieve a Project Milestone within 12 months of the specified date not due to Force Majeure or Excused Delays shall be an Event of Default by Seller.

(continued)

TABLE A.1 Continued

Capacity Transfer	Seller will transfer physical capacity to Buyer starting with capacity to be delivered in 2012/2013 planning year and annually thereafter via PJM eRPM system. Buyer will be responsible for bidding the capacity into the PJM RPM Auctions.
Capacity Payment (2012 dollars)	$___ per kW-mo (2012 dollars) multiplied by the fixed contract capacity of 640 mW ("Contract Capacity"). The capacity payment will consist of two components: Capacity Charge: $___ per kW-mo [flat (i.e., not escalating)] multiplied by Contract Capacity Fixed O&M Charge: $___ per kW-mo escalating annually at CPI multiplied by Contract Capacity CPI multiplied by Contract Capacity
Quantity	Energy offered to Buyer will be available to be scheduled based on an Annual Capacity Test and adjusted for temperature and humidity conditions for each hour during the year. The following quantities are the expected new and clean outputs at various temperatures: 92F Base Quantity 517 mW 59F Base Quantity 545 mW 10F Base Quantity 565 mW 92F Duct Burner Quantity 113 mW 59F Duct Burner Quantity 115 mW 10F Duct Burner Quantity 109 mW
Contract Heat Rates	The Hourly Base Load Heat Rate and the Hourly Duct Burner Heat Rate used for the Energy Payment will be based on the annual heat rate test, which shall be conducted annually including the first year, and adjusted for actual hourly temperature and humidity conditions during the year. The following are the expected new and clean heat rates at ISO conditions. Base Load Heat Rate 6.70 MMBtu/mWh Duct Burner Heat Rate 9.0 MMBtu/mWh
Financial Energy Scheduling	During the Term and subject to the provisions hereof, Buyer may elect to schedule its Financial, Day-Ahead Energy under the provisions set out in clauses (i), (ii) and (iii) below. (i) Quantity — Buyer may schedule for each hour either: (a) No Energy, (b) the Base Quantity, or (c) the Base Quantity plus the Duct Burner Quantity. (ii) Hours of Scheduled Operation: (a) The Base Quantity shall be scheduled for no fewer than six consecutive hours.

(b) The Duct Burner Quantity shall be scheduled only when the Base Quantity has also been scheduled and may be scheduled for any hour or hours contiguous or non-contiguous within any block permitted under the foregoing clause (a) without incurring a Start Charge.

(c) Scheduling the Facility will be limited to operating restrictions of the Facility, including but not limited to restrictions in the Air Permit that limit the annual # of starts per year to approximately 200 factored starts.

(d) Following the last hour of a schedule, the next scheduled hour cannot be incurred until six hours have elapsed, i.e., minimum down time of six hours.

(iii) Daily Scheduling: For each calendar day, Buyer shall notify Seller via an Excel spreadsheet in an e-mail the quantity that it is scheduling for each hour for such calendar day by 8:00 a.m. EPT on the Gas Business Day immediately prior to such calendar day.

Energy Payment and Financial Settlement Procedure

Seller pays Buyer the sum of all hourly Energy Payments (defined below) over the applicable Payment Period plus the aggregate Start Charges for such Payment Period.

For each hour of exercise (excluding Forced Outage hours), the Energy Payment shall equal:

$$[EIP - [((GIP + GA) * CHR) + VOM]] * NQ$$

For the purposes of the above calculation:

"EIP" means Electricity Index Price

"GIP" means Gas Index Price

"GA" means Gas Adder

"CHR" means Contract Heat Rate which shall be the weighted average of the Hourly Base Load Heat Rate and Duct Burner Heat Rate

"NQ" means Notional Quality, which shall equal, as applicable, the total Quantity exercised for each hour

"VOM" means Variable O&M Charge. See the Major Maintenance Charge section

Start Charge: See the Major Maintenance Charge section. A Start Charge will apply based on the Hour/Start Ratio and it will be based on the number of scheduled starts over a Payment Period. Start Charges will be incurred based on the schedule not actual starts of the facility.

Outages: In the event that there is a Forced Outage or Scheduled Maintenance for a particular hour, there will

(continued)

TABLE A.1   Continued

	not be a financial settlement for the Energy Payment for that particular hour, i.e., the Transaction is unit contingent.
Gas Index Price	The price per MMBtu of Gas stated in U.S. dollars and published under the heading "Daily Price Survey ($/MMBtu): Citygates:
	Transco, Zone 6 Non-NY: Midpoint" for the flow date that
	Corresponds to the Scheduled Energy, as published by Platts, a Division of the McGraw-Hill Companies Inc. in the Daily Price Survey Section of Gas Daily.
Electricity Index Price	For the applicable hour, the Hourly, Day Ahead Market LMP Nodal price for the Facility (PJM PNode), as published by PJM.
Gas Adder	$0.13 per MMBtu November through March
	$0.02 per MMBtu April through October
Major Maintenance Charge	The Major Maintenance Charge is intended to be a pass through Seller's variable operating costs. The Seller does not intend to over collect or under collect these variable operating costs. The Major Maintenance Charge is a combination of a Start Charge and a VOM Charge, which depend on the ratio of scheduled starts to operating hours between maintenance intervals.
	The below Major Maintenance Charges (2012 dollars) shall escalate annually at CPI. They are based on a preliminary GE long term service agreement proposal for a similar facility.

Hours/Start Ratio	< 26:1	> 26:1
Start Charge ($/Start)	$30,400 + (3,000 MMBtu * Gas Index)	$0 + (3,000 MMBtu * Gas Index)
VOM ($/MWh)	$.90/MWh	$2.85/MWh

Each month, the Project will charge the Buyer a Start Charge and a VOM Charge pursuant to that month's ratio of hours to start based on the number of scheduled starts and number of scheduled hours. At the end of a maintenance interval, Seller shall true up any difference in monthly Start Charges and VOM Charges paid by Buyer with the corrected Start Charges and VOM

	Charges using the actual ratio of hours to starts between maintenance intervals. A maintenance interval will occur every 16,000 operating hours or 375 starts, whichever is sooner.
Availability	This Transaction is unit contingent. However, Seller will guarantee that the Facility's Forced Outage Rate (as defined by GADS) will not be greater than 5% on an annual basis (the Guaranteed Forced Outage Rate).

Buyer's sole remedy for Seller not meeting the Guarantee Forced Outage Rate will be based on the following:

In the event that the Facility's Forced Outage Rate is higher than the Guaranteed Forced Outage Rate, the Annual Capacity Charge shall be reduced by 1 percentage point for each 1 percentage point of increased Forced Outage Rate.

In the event the Facility's Forced Outage Rate is less than the Guaranteed Forced Outage Rate, the Annual Capacity Charge shall be increased by 1 percentage point for each 1 percentage point of decreased Forced Outage Rate up to a decrease of two percentage points.

Scheduled Maintenance hours will not be considered Forced Outage hours. Scheduled Maintenance will be based on equipment manufacturer recommendations. Seller shall provide Buyer reasonable notice of Scheduled Maintenance periods.

In the event that there is a Forced Outage or Scheduled Maintenance for a particular hour, there will not be a financial settlement for the Energy Payment for that particular hour, i.e. the Transaction is unit contingent.

Scheduling/Fuel Procurement/ Transportation	Seller shall be responsible for all scheduling in compliance with PJM protocols.
	Seller shall be responsible for the management, procurement and transportation of all natural gas for the Project.
Carbon Costs	Buyer shall be responsible for the costs of securing all required carbon allowances and shall retain all rights to the benefits of the carbon credits or allowances available to the Project. The Project may be eligible for up to 100% of the CO_2 allowances necessary to operate the plant for up to the first six years of operation, pursuant to MDE's Clean Generation Set Aside of the State RGGI program.

DTC's Coal vs. Natgas Displacement Model Methodology, January 6, 2009

This material here provides a comparison between natural gas and coal fired power plants from a report by Doyle Trading Consultants, LLC. It reviews the factors that are required to complete this analysis and reviews some of challenges of shifting from coal to natural gas power plants. This analysis is important for renewable power plants since they operate in a world of both coal and natural gas–fired power plants.

DTC'S COAL/NATGAS DISPLACEMENT MODEL METHODOLOGY

Executive Summary: Congratulations! If you are reading this report, then you realize that it is sheer folly to hope to derive a truly digital coal vs. natgas "switching price" from something as complicated as our nation's grid. So why did DTC invest thousands of dollars and hundreds of hours into a very complicated model that, in fact, results in a digital "switching price"? Obviously, we believe that there is a level to where natgas prices can drop, which will result in a meaningful displacement of coal-fired generation. However, we do not believe that our model delineates the "switching point." Instead, we believe it shows that at some level *below* that switching price, material displacement could occur. We believe that if the user is acquainted with our methodology and is armed with very important caveats, he/she can use the output as a tool to gauge the vulnerability (or lack thereof) that coal-fired plants have to being displaced by natgas plants. This report provides an overview of our methodology and, more important, provides some very important color around the factors that determines whether natural gas plants dispatch ahead of coal plants. See Figure B.1.

		DTC's Coal vs. Natgas Model© (January 6, 2009)				
		10 Heat Rate Coal Plant (without SCR or Scrubber) vs. 7 Heat Rate Natgas Plant				
		Model Comparison Period: Q2 2009 (All-in Costs per MWh)				
		Inputs: Current Coal, Emissions, Natgas Prices				
NERC Region	PJM East	PJM West	NYISO West	Ill (Midwest)	ERCOT	SERC
Coal Source	Napp 13000 3.0# rail	Napp 13000 3.0# rail	Napp 13000 3.0# rail	Ill. Basin 11800 2# Barge	PRB 8800	Capp CSX 12500 1%
Coal Gen Costs ($/MWh)	$50.12	$45.32	$50.46	$41.98	$33.49	$46.20
Natgas Gen Costs ($/MWh)	$50.10	$48.10	$50.10	$48.10	$48.10	$48.10
Spread ($/MWh)	($0.03)	$2.78	($0.36)	$6.13	$14.61	$1.90
Natgas Breakeven ($/MMBtu)	$6.91	$6.22	$6.96	$5.75	$4.53	$6.35

Note 1: Negative spread signifies the natgas gen costs are *theoretically* cheaper than coal gen costs.
Note 2: For information on DTC's methodology and some very important caveats on interpreting the model's output, please contact Ted O'Brien (646.840.1300; tobrien@doyletradingconsultants.com) for our detailed 'white paper' on this subject.

FIGURE B.1

DTC's Methodology

■ **Time period:** Our model calculates the switching price based on prompt quarter. We also calculate prompt year for in-house purposes, but do not include the results in our daily flash e-mail. We do not use prompt month, due to the fact that most coal-fired power plants are not yet nimble enough to react to prompt month market signals. Coal and transportation logistics are almost always scheduled at least one month in advance. While a genco could make the decision to take a coal plant offline and dispatch natgas, it would be doing this to conserve inventory and not necessarily for economic reasons. *Note:* Some merchant generators are shifting to prompt month, but the majority of the grid has not yet adapted this.

■ **NERC power regions:** Due to several factors, determining a nation-wide, one-size-fits-all switching price is implausible. Therefore, we chose six power regions for our model: NY ISO West, PJM East, PJM West, SERC, Illinois/Midwest, and Ercot.

■ **Generic power plant:** We chose a generic power plant for each region that uses the most common coal quality for that region. In the real world, some power plants are dependent on a special coal quality (e.g., Norfolk Southern compliance coal), which puts that specific plant at a higher risk of being displaced by natgas. Conversely, there are some plants within the same region that might be very close to the coal mines or perhaps have excellent transportation alternatives and much lower transportation costs (and are therefore at a lower risk of being displaced by natgas). We chose a middle-of-the-road generic power plant for our model.

■ **Minimum emission controls:** Our goal is to find the threshold price where material, incremental displacement occurs. While we use heat rates (see below) that are considered reasonably efficient, our model assumes that the power plants do *not* have the most sophisticated emissions controls (scrubbers for sulfur dioxide [SO_2] removal; selective catalytic reduction [SCR] for nitrous oxide [NOx] removal). This is an important issue since SO_2 and NOx allowances can result in an additional +$1.50/MMBtu for a coal plant with such controls. However, if we were only to model such plants, we would miss our objective of identifying at what point material and incremental displacement occurs.

■ **Heat rates:**
  ■ **Coal plants:** The most efficient coal plants use about 9.5 MMBtu to produce 1 megawatt hour (mWh) of electricity; the least efficient coal plants use about 12.5–13.0 MMBtu to produce 1 mWh. For our model, we use a 10 heat rate (10 MMBtu = 1 mWh).

- **Natgas plants:** The most efficient natgas plants use about 6.6 MMBtu to produce 1 mWh; the least efficient natgas plants use as much as 15 MMBtu to produce 1 mWh. For our model, we use a 7 heat rate (7 MMBtu = 1 mWh).
- **Coal prices:** DTC uses the end of day settle prices from the OTC market for the prompt quarter and prompt year.
- **Coal quality:** DTC uses the most common coal quality for each region. In virtually every region, power plants consume coal qualities ranging from high sulfur to low sulfur; from high ash to low ash; from bituminous to sub-bituminous. The exact quality and/or blend of qualities can have a direct impact on how vulnerable (or impervious) the coal-fired plant is to displacement by natgas.
- **Natgas prices:** DTC uses the Nymex Henry Hub prices for the prompt quarter and prompt year.
- **Natgas pipeline costs:** DTC uses a fixed $0.50/MMBtu to capture the pipeline cost for natgas in all power regions. Pipeline fees are market-driven and can fluctuate according to supply and demand. We are unable to capture this.
- **Emission prices:**
  - **SO_2:** Our model uses prices for SO_2 allowances based on end-of-day settle prices in the over-the-counter (OTC) market.
  - **NOx:** Most states east of the Mississippi and those western states bordering the Mississippi must comply with NOx emissions during the May–September "ozone" season. We use prices derived from the OTC market. On December 23, an important court ruling reinstated the EPA's Clean Air Interstate Rule (CAIR), and it will remain in effect until a new rule is formed to replace it. CAIR comprises the same group of states that comply to the ozone season rules (May–September) and also includes Texas. Ozone season NOx credits have always been included in our model. Since the CAIR ruling, we have added the cost of CAIR annual NOx credits into our model. The difference in prices is dramatic: ozone season NOx credits in early January 2009 were trading at 675 (approx. $0.15/MMBtu impact), whereas annual NOx credits were trading at $5,250 (approx. $1.20/MMBtu impact). It is important to recognize that during the ozone season, affected states must comply with ozone season NOx credits and CAIR annual NOx credits.
  - **Mercury:** Eventually, there will be a nationwide compliance regulation for mercury emissions and presumably a cap-and-trade program from which emission prices will be derived. We have included a model input for this feature, but it is currently entered in as "zero."
  - **Carbon dioxide (CO_2):** The Regional Greenhouse Gas Initiative (RGGI) was enacted January 1, 2009, which causes the cost of

CO_2 credits to be added at both coal- and natgas-burning utilities in RGGI states (Maine, New Hampshire, Vermont, Connecticut, New York, New Jersey, Delaware, Masssachusetts, Maryland, Rhode Island). While the cost of the credits per megawatt hour varies depending on the characteristics of the individual plant (heat rate, efficiency, etc.) as well as the price of the credits, coal plants typically incur double the cost per megawatt hour as compared natgas plants due to RGGI. We include only RGGI CO_2 credits to NY ISO West and PJM East. *Note:* Even though Pennsylvania plants are included in PJM East and Pennsylvania is only an "observer" to RGGI,, we include RGGI, CO_2 credits to PJM East since Maryland, Delaware, and New Jersey are RGGI states and have significant coal generation.

- **Transportation prices:** DTC uses our best guess of transportation rates as if the genco recently negotiated the contract and will be using train sets owned by the genco. The model assumes that the coal plant is served by captive rail service (only one rail company has access to the plant); approximately two thirds of utilities are served by captive rail; gencos that are served by dual railroads typically have lower transportation rates.
- **Fuel surcharge prices:** We estimate fuel surcharges that would be applied to our transportation rates that are based on fuel surcharge clauses that are prevalent.
- **Operations and maintenance (O&M) costs:** We used representative O&M costs for natgas- coal-fired plants. O&M costs are very sensitive to the type of pollution equipment being used at the power plants.
- **Power prices:** We do not incorporate power prices into our model, but keep in mind that when power prices (especially on-peak) are *above* the generation costs of our generic coal-fired and natgas-powered plants, then both plants will be dispatched onto the grid, regardless of which one is lowest in cost.
- **Natgas heat rates and capacity availability:** For initial displacement to occur there must be sufficient capacity of underutilized efficient natgas plants available to displace coal plants. For significant displacement to occur, less efficient natgas will have to be used (i.e., we will have to go further up the dispatch stack to natgas plants with 7.5 heat rates, 8.0 heat rates, etc.). Obviously, the higher the heat rate of the natgas plant, the higher the natgas switching price will be. In the real world, the least efficient coal plants (12 heat rates) will be the first to be displaced by natgas. And for a significant displacement to occur, more efficient coal plants will have to be displaced by natgas

generation (i.e., we will have to go further down the dispatch stack
to coal-fired plants with 11.5 heat rates, 12 heat rates, etc.).

Grid Reliability and Transmission Issues

■ **Grid and voltage stability:** Some coal plants might be deemed "must-
run" for reasons of grid reliability and cannot be economically displaced by natgas.
■ **Transmission limitations:** Due to infrastructure, there may be limitations on the amount of energy that can be transmitted through power
lines coming from certain plants.

Operational Issues

■ **Start-up risks:** Once a coal plant is shut down, there is a serious risk that
the start-up will not go smoothly (tube leaks, turbine vibration, etc.) and
the plant will not be available when expected to resume operations. Plant
managers are loath to allow coal-fired plants to go "cold" for short-term
economic reasons. Therefore, the decision to displace coal-fired generation would have to entail an extended, several-day period and not just a
several-hour period in between on-peak and off-peak periods.
■ **Coal plant start-up costs:** To restart a coal plant, between 1,000 and
2,500 barrels of #2 oil can be needed (with 42 gallons in each barrel).
We do not model this, but it is a cost that is factored in by the gencos.
■ **Natural gas plant start-up costs:** It can take up to 4 to 6 hours to start a
natural gas plant, with between 17 and 25 percent of the daily gas requirements expended to get the plant to the necessary temperature before generating electricity. We do not model this, but it is a cost that is
factored in by the gencos.
■ **Heat rate penalty:** A decision could be made to ramp-down a coal
plant to minimal output to allow for short-term economic displacement. However, as a rule of thumb, a plant manager would not
want to ramp-down a 600-mW plant below 300 mW for reliability
issues. The efficiency of a 600-mW plant with a 10 heat rate can
drop to a 12 heat rate when ramped down to 300 mW. (Turbines in
combined-cycle gas plants also have operational limitations, and
plant managers have a difficult time running below 75 to 80 percent
of capacity. Many installations have multiple turbines, in which case
the plant manager can opt to shut down one turbine while keeping
the others at near-max capacity.)
■ **Warranty issues:** In order to maintain manufacturer warranties, major
multimillion-dollar overhauls requiring multiple-week outages are

mandatory after a certain number of operating hours and/or cold starts. Warranty issues are an important input in the coal vs. natgas decision matrix.

- **Natgas commitments:** We understand from a genco expert that a natgas plant will not replace a coal plant without purchasing at least a few days of natgas consumption (as a precaution for a delayed restart of the coal plant). We do not model this, but it is a cost that is factored in by the gencos.
- **Pipeline capacity:** While the economics could argue for widespread displacement of coal-fired plants by natgas plants, the availability of pipeline capacity could be a constraining factor in coal-heavy electricity regions.
- **Inventory costs:** Most power plants prefer to unload coal directly into the power plant bunkers. An additional cost of $1.50/ton is incurred when the coal is sent to the stockpile.
- **Inventory constraints:** Whereas the model might argue for the natgas plant to dispatch ahead of the coal plant (and allow the genco to ship coal into inventory), some gencos are constrained by a certain stockpile footprint above which costs can be prohibitive.
- **Take-or-pay coal contracts:** Some generators have take-or-pay coal contracts and are unable to participate in economic dispatch of natgas generation.
- **Liquidated damages for transportation shortfall:** Under some transportation agreements, penalties of $2.00 to $3.50 per ton can be incurred if coal is not delivered pursuant to the agreement.
- **Self-correcting market:** Marginal displacement will have an impact on coal, natgas, and emission prices; therefore, the displacement "window" may be for only a short time frame. Ninety percent of U. S. coal production is used for power generation compared to approximately 32 percent of natgas production that is used for power generation. This results in natgas prices being more sensitive to incremental changes in demand.
- **On-peak/off-peak power prices:** Power prices fluctuate according to demand, and gas generation usually sets the prices. In most cases, on-peak power prices are high enough to dispatch *all* coal generation and efficient gas generation. Most displacement will occur during off-peak hours.
- **Inability to resell coal:** Many gencos cannot resell coal in order to pursue economic displacement. Some are prohibited by specific contractual clauses; some have "understandings" that they won't resell the coal; some have no in-house procedures to resell coal.

Regulated Genco Issues

- **Market-priced coal and transportation contracts:** The model incorporates market prices for coal and transportation. In the case of coal, virtually all utilities have a mix of spot and term contracts, which can be as much as 50 percent below prevailing market prices. In the case of rail, many gencos have legacy, below-market rates that are 30 to 40 percent below prevailing market prices. Most regulated gencos (and many unregulated gencos) dispatch coal plants based on average cost of delivered coal, which can dilute the relevancy of our model's switching price.
- **Emissions allowances:** Most utilities are granted a certain number of emission allowances at zero cost by the Environmental Protection Agency (EPA). Although many utilities calculate emission allowances at market when determining dispatch, a surprisingly high number still use a zero price for the allowances that they have been allocated (making coal generation more competitive as compared to natgas).
- **Utility carrying costs:** The public utility commissions typically allot the regulated utility a certain number of days of coal inventory, above which the carrying costs are borne strictly by the utility, not ratepayers. (All carrying costs of inventory are borne by merchant gencos.)
- **Merchant gencos:** About 30 percent of the nation's coal-fired fleet is in the hands of merchant generators. These entities are more likely to dispatch based on market-to-market price signals. However, they are still confined by many of the same issues as the regulated generators (take-or-pay contracts for coal and rail, inability to resell some coal, etc.). Merchant gencos are more apt to take risks from which they will profit (as opposed to their regulated counterparts, whose shareholders would have to bear the risk and hand over the profit to the ratepayers).

HOW MUCH NATGAS IS NEEDED TO DISPLACE COAL?

Many clients are interested in knowing how much natgas is needed to replace a certain monthly volume of coal. The exact volume varies according to coal quality and to the efficiency of the natgas unit that is doing the displacement. In our table, we chose the Capp coal (12,500 btu/lb.) since Capp coal is usually on the displacement firing line. We chose two natgas heat rates to provide an idea for the magnitude of natgas consumption when generators are forced to move up the dispatch stack to use less efficient gas units to displace coal generation. See Figure B.2.

DTC Monthly Coal/NatGas Displacement Tool			
Monthly NatGas Required to Displace Coal			
Tons of CAPP Coal Displaced (mm/month)	MW Capacity	BCF NatGas Required (7 Heat Rate)	BCF NatGas Required (9 Heat Rate)
1	3,425	17.5	22.5
2	6,849	35.1	45.0
3	10,274	52.6	67.5
4	13,699	70.2	90.0
5	17,123	87.7	112.5
10	34,247	175.5	225.0
15	51,370	263.2	337.5
20	68,493	351.0	450.0
25	85,616	438.7	562.5
30	102,740	526.4	675.0
Doyle Trading Consultants, LLC (970 256 1194/1192)			

FIGURE B.2

About the Authors

Tom Fogarty has spent his entire over-25-year career in the energy business and is an energy executive at a major international corporation. Prior to this, he founded and led PNT Energy, an energy restructuring and investment consulting practice to early- and late-stage private equity investors, corporations, hedge funds, and developers. He has an extensive international background in the development, financing, technology, design, valuation, operations, and restructuring of gas, nuclear, wind, solar, hydro, geothermal, landfill gas, biomass, coal, and waste coal electric power assets. Prior to PNT, Tom was with Sithe Energy and Foster Wheeler. He has also written an editorial in the *Daily Bankruptcy Review* and has been quoted in other sources on the many current challenges facing renewable power. He received his MBA from New York University Stern School of Business and his BSME from Fairfield University.

Robert Lamb is currently a professor at New York University's Stern School of Business and a management consultant. He was previously strategy adviser and debt adviser to New York State Power Authority, and over the past 25 years, he has developed and taught a number of customized courses for specific investment banks and corporations including Goldman Sachs, Deutsche Bank, Merrill Lynch, Morgan Stanley, and Citibank American Express. He has written books and chapters on the financing of public power projects. He is also a founding member of Standard & Poor's Academic Counsel of Advisors.